YO-EAZ-962

ELECTRICAL CONTRACTING

Electrical Contracting

RAY ASHLEY

Business and Estimating Consultant—Electrical Construction
Consulting Editor of *Electrical Construction and Maintenance*
Registered Professional Engineer
Member, Chicago Electrical Estimators' Association

McGRAW-HILL BOOK COMPANY, INC.

New York   Toronto   London   1961

Copyright © 1961 by the McGraw-Hill Book Company, Inc. Printed in the United States of America. All rights reserved. This book, or parts thereof, may not be reproduced in any form without permission of the publishers. *Library of Congress Catalog Card Number: 61-7303*

ISBN 07-002408-1

10 11 12 13 14 15 HDBP 79876

*I*ntroduction

Additional margins of markup, gains by selecting better contracts, reducing labor cost by good management, obtaining new customers, and many other benefits are subjects of this book.

To the man who dreams of big stakes, these seem like small items of business, but in everyday electrical contracting, competition is so keen and the barriers between success and failure so thin that the average electrical contractor cannot afford to ignore them.

Although the first few chapters are written principally for the prospective electrical contractor, the main text is vitally concerned with the affairs of current businesses. The problems of knowing how to avoid overmanning jobs, how to keep work from dragging, and how to avoid falling victim to the slow progress of other trades are all items of labor management as important to existing contractors as to the newcomer. Again, the old as well as the new contractor must learn to select better work and to make the best of contracts he already has. Benefits realized from everyday gains on the average contract far outweigh those from the occasional "killing" on a single project.

Throughout this book various wage rates and methods of markup have been employed to indicate to the reader that the studies do not necessarily apply to any particular year, locality, or scale of wages. Rather, they are intended to illustrate stable methods and practices suitable to any time and all communities. In many cases, the reader must work up his own studies by substituting wages and markups suitable to his locality and type of business.

Over twenty-five years' experience in all phases of the electrical contracting industry has thoroughly acquainted the author with the problems involved. Constant research and contact with contractors and estimators in all parts of the United States and Canada have made it possible to offer solutions to many of the problems.

The author wishes to acknowledge the kindness and cooperation of members of the Chicago Electrical Estimators' Association, Chicago contractors, and many individuals throughout the United States and Canada for assistance in promoting research work. Among individuals promoting research in the electrical contracting industry, the work of Walter Brand, Los Angeles, California, has been outstanding.

Ray Ashley

Contents

Introduction v

List of Illustrations xi

1. *So You Think You Ought to Be an Electrical Contractor?* 1

   Careful study versus wishful thinking. Mistaken ideas about personal qualifications. Reasons for going into business. Some flimsy reasons. Hazards. Business failures. Competition encountered. It is a good business. Confidence and enthusiasm essential.

2. *Do You Know How Much It Costs to Get Started in Business?* . . . 9

   Expenses to be considered. A study of a specific type of business. Division of proprietor's time. Office equipment and costs. Tool costs. Office location and rents.

3. *Do You Know How Much It Will Cost You per Month While You Are Waiting for Contracts?* 23

   Salaries cannot be ignored. The first six months' operating expenses. Estimated contracts for the first six months. Expenses will exceed income. Time passes quickly while waiting for contracts.

4. *Do You Know How Much Business Must Be Forthcoming to Pay Operating Expenses?* 29

   $100,000 with liberal markups—the minimum required. Direct job costs and overhead. Estimated operating costs (minimum) for $100,000 volume. Effects of high operating costs and low markups.

5. *Do You Know Where You Are Going to Get Contracts for $100,000?* . . 33

   The mistaken idea that contracts come automatically. Business obtained from established connections necessary. Contracts must be assured. Estimating time a limiting factor. Labor burden approximately $4,700 per mechanic.

6. *Do You Know Where You Are Going to Get the Men?* 41

   Men must be good if they are to meet competition and satisfy customers. Manpower is one of contractors' biggest problems. An average of five men needed. Classification of mechanics. Apprentice training. Competing with established contractors for men. Good mechanics want to work for established contractors. Cost of inferior mechanics. Proprietors do not work with tools.

7. *Are You Familiar with the Functions of Contracting?* 49

   Contractors must have a good knowledge of estimating, engineering, and how to direct men. Hiring good estimators may not solve the estimating problems. Resourcefulness and ingenuity required in competition. The better engineer has the advantage.

8. *Can You Select the Better Contracts?* 61

   Volume depends on type of work. Hazards of various types of work. Desirable contracts. Future business. Repair services. A case of good judgment.

9. *Can You Make the Most of Contracts?* 71

   Management problems numerous. Selecting mechanics and organizing crews. Directing work. The need of construction drawings. Using tools to the best advantage. Proper and timely delivery of materials. The contractor must maintain control of his work.

10. *Are You Familiar with Operating Costs?* 81

    A contractor must understand his own operating costs. Definitions of operating costs. Billing direct job costs (DJC). Profit a definite gain. Why profit disappears. Surveys by others misleading. Studies serve as guides only.

11. *Do You Know How to Conduct Overhead Studies?* 89

    Errors made in estimating overhead. Fatuous values used in many offices. Reliable overhead studies provided by others useful. Preliminary job studies. Preliminary overhead studies. Complete-volume studies. Adjusting unbalanced overheads. Material services (installation-only project). Value of limited studies.

12. *Can You Estimate Installation-only Contracts?* 103

    Definition of installation-only (I-O) contracts. Erroneously called "labor-only" jobs. I-O services. I-O hazards. Excess labor. Three types of I-O contracts. Relative costs of supplying labor and material. Listing direct job expenses. The estimate as a selling aid.

13. *Can You Write Proposals?* 119

    Bidding according to plans and specifications. The sample proposal. Supplying completion dates. Dangerous practices. Fixed-price contracts, no plans. Cost-plus bidding. Forms of cost-plus bidding. Bidding according to the hourly rate of pay. Useful information to be supplied. Facsimile of a special proposal. Treatment of verbal orders.

Contents　　　　　　　　　　　　　　　　　　　　　　　　　　ix

14. *What Type of Organization Do You Need?* 133

Sole owner and proprietorship. Partnership. Limited partnership. Corporation. The individual with his business incorporated.

15. *Can You Collect?* 141

Six types of contracts to be billed. The small fixed-price contract. The small cost-plus contract. The cost-plus contract with no stipulated markups. The large fixed-price contract to be billed as work progresses. Establishing prices for large fixed-fee contracts. Preparing to make draws on large fixed-price contracts. Preparing the divisions of contract. Amounts withheld. Requesting final payment. Wavers of lien and lien rights.

16. *Can You Conduct Labor-cost Studies?* 153

Labor-cost studies a must item. Planned programs required. Breakdown (time and motion) studies. Job-division studies. Prorata labor for completed projects. Using labor-cost studies to adjust estimating units. Analysis of completed projects. Studying completed projects that have lost money. Mistakes of management revealed.

17. *Do You Understand Incidental Costs?* 177

"Nonproductive labor" a misnomer. Use of better terms. Job labor and incidental labor. Incidental labor included in standard labor unit. Incidental labor and direct job costs. Incidental labor check list for estimating and selling. Incidental labor not overhead. Billing incidental labor. Satisfying buyers' demands for cost analysis and detailed costs.

18. *Do You Know What It Costs to Train Apprentices?* 193

Necessity for training apprentices. Division of costs. Productive efficiency of apprentices. Expense of training. Who bears the burden and who benefits.

19. *Can You Explain Your Tool Charges?* 201

Factors affecting tool costs. Time in use versus rated life. Fixed costs. Individual job expense. Billing for tools.

20. *Do You Understand Your Labor Curves?* 209

Man-power demand curves a part of good business. Labor curves reflect the type of management.

21. *Watch Your Trim Labor* 215

Effects of management faults on trim labor. Common use of the wrong labor units. Psychological conditions affecting trim labor. Faulty appraisals by contractors. Necessity of reducing crews at the right time.

22. *The Wide Bid Spread* 221

Wide bid spreads do not necessarily indicate "wild bidding." Causes for wide variations in sound bids.

23. *Safe Billing* 229

Comprehensive billing. Care required in billing regular accounts. Adjusting overhead to suit the type as well as the size of the contract. Separate markups get better acceptance. The demoralizing effects of losing regular customers.

24. *Architects and Engineers* 237

Cautions to be observed when dealing with architects and engineers. Methods of billing change orders and "extras." Unit-price bidding. Various other types of bidding. Opportunities for educational work. Care required in recommending changes. Protecting the architect and owner.

25. *The Estimator* 247

Minimum requirements. Management and estimating problems. Freedom of activity. Responsibilities of estimators. Overtime pay for estimators. Getting bids out on time.

26. *Man-power Productivity (MPP)* 253

Factors affecting MPP. The labor market. Qualifications of the individual mechanic. Attitude of mechanics. A study of incidental and preparation time. Crew productivity. Effects of working conditions.

27. *Advertising* 263

Types of advertising used. Direct contact considered the best form. Use of signs on trucks. Limited use of newspapers. Brochures and their limitations. Employing expert advertising men. Miscellaneous forms. Expense allowances. Advertising budget.

28. *Continuing in Business* 269

Contractors cannot relax their hold. Impatience and depressions are serious hazards. Important decisions of management. Building up a reserve. $18,000 reserve needed for a one-man organization. Allowance for reserve: 3%. Contractors' pride in their business.

Definitions 275
Abbreviations 279
Index 281

List of Illustrations

Fig. No.	Title	Page
1-1	Office Layout	5
2-1	Division of Proprietor's Time	10
2-2	Estimate of Furniture Costs	11
2-3	Tools and Equipment for a Five-man Crew	13
2-4	Tool Check List	14
2-5	Rolling Scaffolds and Platform Hoist	15
2-6	Band Saw	16
2-7	Hydraulic Pipe Bender	16
2-8	Platform Hoist	17
2-9	Semitrailer	17
2-10	Truck with Hydraulic Platform Lift	18
2-11	Panel Truck	18
2-12	Special Truck with Overhead Pipe and Ladder Rack	19
2-13	Material Check List	20
3-1	Operating Costs for a One-man Business	24
3-2	First Six Months' Contracts	25
3-3	Plan for Office Expansion	27
4-1	Operating Costs—$100,000 Volume	30
4-2	Distribution of Costs—$100,000 Volume	31
5-1	Labor Burden—Job Costs per Mechanic Employed	37
7-1	Factory Lighting—Outlets Located Only	53
7-2	Factory Layout—Contractor's Working Drawings	54
7-3	Factory Power—Motor Sizes and Location Only	56
7-4	Factory Power—Contractor's Working Drawings	57
7-5	Factory Power—Contractor's "Riser Diagram"	58
7-6	Contractor's Take-off—Power	59
8-1	Optimum Duration Tables	66
8-2	Repair-service—Operating Costs	68
9-1	Shop Drawing—Feeder Conduit Entering Shaft	74
9-2	Shop Drawing	76
9-3	Effect of Good Management	78–79
10-1	A Business Reminder—"Profit Disappears"	83
10-2	Early Surveys of Operating Costs	85
10-3	Operating Costs—Direct Job Expense and Overhead	86
10-4	Operating Costs—Various Types of Work	88
11-1	Overhead Curves—$100 to $9,000 Projects	91
11-2	Overhead Curves—$10,000 to $100,000	92
11-3	Overhead Curves—$100,000 to $1,000,000	93

List of Illustrations

Fig. No.	Title	Page
11-4	Preliminary Check of Annual Overhead	95
11-5	Overhead Study for a Year's Business	96
12-1	Excess Burden—Installation-only Projects	105
13-1	Proposition for Electrical Work	121
13-2	Analysis of Labor Costs	126
13-3	A Proposition for Electrical Work (Special)	127
13-4	A Preliminary Estimate (a Special Case)	130
15-1	Cost-plus Billing	143
15-2	Cost-plus Billing—Alternate Method	145
15-3	Cost-plus Billing—Second Alternate	146
15-4	Division of Costs for an Electrical Contract	148
15-5	An Estimate of Completed Work	149
15-6	A Final Waiver of Lien	152
16-1	Outlet Boxes—Breakdown Study (Time and Motion)	155
16-2	Elbows—Breakdown Study of Labor	156
16-3	Outlet Box—Pictorial Study	157
16-4	Pipe Entrances—Pictorial Study	158
16-5	Cable-pulling Records (Field)	160
16-6	Cable-installation Data	161
16-6A	Cable-installation Records (Tabulated)	162
16-7	Conduit-installation Data—Group Runs	163
16-8	Conduit-installation Data—Single Runs	164
16-9	Job-division Studies—Installation Time	166
16-10	Job-division Studies—Installation Hours	168
16-11	Job Time Study—Branch-circuit Conduit	169
16-11A	Job Time Study—Branch Circuit and Feeders	170
17-1	Standard Labor Unit—Breakdown 3-in. Conduit	179
17-2	Check List—Incidental Labor	181
17-3	Direct Job Costs—Special Study Form	183
17-4	Overhead Expense—Special Study Form	183
17-5	Percentage Method of Figuring Incidental Costs	184
17-6	Breakdown—Direct-job-cost Percentages	185
17-7	Incidental Cost Estimate—Listing Method	187
17-8	Summary and Bid Sheet	188
17-9	Analysis of Labor Costs	190
17-10	Sliding Scales of Costs	191
18-1	Apprentice-training Costs	195
19-1	Estimated Cost of Tools—50 Electricians	205
19-2	Estimated Cost of Tools—5 Electricians	206
20-1	Labor Curves—Neglected Job	210
20-2	Labor Curves—Special Peak Demands	211
20-3	Labor Curves—Showing Effect of Management	213
21-1	Job Progress Followed Closely	217
21-2	Careful Management Planning Necessary	218
22-1	A Bid below Normal—Justified	223
22-2	A Bid above Normal—Justified	224
22-3	A Normal Bid	225
22-4	Wide Spread in Bids Due to Markup	227
23-1	Overhead Curves	231
23-2	A Study of Overhead—for Regular Customers	232
23-3	Adjusted Markups	234
26-1	Attitude of Mechanics—a Study	257
27-1	Page from a Brochure	267

CHAPTER 1

So You Think You Ought to Be an Electrical Contractor?

So you think you ought to go into the electrical contracting business? And what makes you think so? Is it because you have proved yourself capable of conducting such a business, or is it because you are giving way to some wishful thinking?

Electrical contracting is a good business, but before going into it, one must be familiar with all the essentials for successful operation. It is reported as being a good business because the better contractors are running it. Most of the poorer ones have failed and passed out of the field.

One may think he has studied the possibilities of electrical contracting carefully, whereas, in reality, he has just been dreaming about it. Constructive thinking, backed up by cold, hard figures, produces a different picture.

MISTAKEN IDEAS

One must not get the mistaken idea that being proficient in one phase of the work is any guarantee of success in the electrical-construction business. Too often, men emphasize their strong points and fail to recognize their weak ones. A man may be a good estimator, foreman, or superintendent but, owing to lack of executive ability and general knowledge of the work, is not qualified to run a business.

One may decide he has executive ability and can hire others to do his estimating, engineering, and superintending. In the first place, he may be overrating his ability, and in the second place, suppose he is a good executive and knows where he is going to get the right men, where is the money coming from to pay such high-priced help?

Some men make up for their shortcomings by taking a partner. This may work out, but right away the business has two heads to support. As we shall see later, partnerships have their disadvantages as well as advantages.

REASONS FOR GOING INTO BUSINESS

Before studying the things to be considered when contemplating electrical contracting, let us look at some of the reasons men have for thinking they should go into business. No doubt the most common is to make money. We all understand that. Two of the main reasons for qualified men going into business for themselves are:

1. To be more independent.
2. There is no alternative.

Many well-trained and well-qualified men find it difficult to shape their work to suit superiors. They want to get out and have their own customers, hire their own help, and own the business.

Some men have no alternative. A contractor develops more top men than the business can support. Someone must cast off and either go to some other contractors or go in business for himself. Many choose the latter.

A similar situation often develops when contractors have sons coming into the office. As sons develop and take over, top men must step aside.

FLIMSY REASONS

One contemplating electrical contracting may do well to study the flimsy reasons that others have had for thinking they should go into business. Such a study will cause him to analyze his own position more carefully.

It is astounding how little it takes to get some men to think they ought to go into the electrical contracting business. The following are a few of the flimsy reasons for such ideas:

1. Someone suggests that they should go into business for themselves.

2. They are tired of working for someone else and want to have someone working for them.
3. Their wives think they should be in business.
4. They want to raise their standard in the community by being in business.
5. Some supply house wants them to go into business.

We cannot dwell long on the foregoing reasons; however, some brief comments are in order.

If anyone ever suggests that you go into business, ask him how much money he is willing to put into it. You will soon know whether or not he has given the matter any serious thought.

Most of us would rather be in business and work for ourselves than work for someone else. That, in itself, does not justify launching out. As we shall see in later chapters, there are many things to be studied before one can decide to elect such independence. Besides, many men would like to be independent but do not care to pay the price by taking risks and responsibilities.

Wives who suggest that their husbands go into business must depend on the husbands to supply the pros and cons on the subject. Many wives, if they were familiar with all details, would hesitate to make such suggestions.

The smaller the community, the more likely one is to want to raise his standing in it by being in business. He should be equally anxious to be prepared to avoid failure. I have long been of the opinion that it is better to be an obscure workman in a community than to be the outstanding bankrupt.

Because a supply house wants a man to go into business is the poorest reason one can think of for his doing it. It is enough here to say that any man who goes into business for that reason alone just is not smart enough to stay in business long.

HAZARDS

The following chapters are designed to acquaint the reader with conditions and requirements to be prepared for when entering the electrical contracting business. After he is prepared, they just represent problems of the business. For one not prepared, these same conditions and requirements represent real hazards.

STRIKES PUT MEN IN BUSINESS

Before dismissing the subject of why men go into business, the reader may be interested in learning about the days when strikes and lockouts were responsible for men going out on their own. Strikes and lockouts were much more common 40 or 50 years ago than they are today. They were often of long duration, and mechanics with families became sorely in need of some source of income.

During the long periods of layoff, many of the men went out and solicited work by the day from factories, commercial institutions, and homes. By the time the labor troubles were settled, some of the mechanics were well established and continued to build up businesses of their own.

In recent years, few men have started in the electrical contracting business as a result of labor troubles. There are reasons for this. In the first place, strikes and lockouts are less common. In the second place, the requirements for getting started in business are much more severe.

If there were labor troubles now, more than likely many of the larger institutions would try to build up their own construction crews. They have handymen who would like to make their jobs bigger by taking care of new installation work.

BUSINESS FAILURES

Failures reported by Dun and Bradstreet and others are evidence enough to prove that many men have started in the electrical contracting business before they were prepared. True, some of the failures may have been due to "bad breaks" or because the business was started at the wrong point in the cycle of a construction period. However, for the most part, it was a case of the wrong man in the electrical-construction business.

Building construction goes in cycles, and one who starts out on the tail end of a cycle may soon find himself without any work. He has a lease, tools, and equipment on his hands, all of which are of no value if there are no contracts.

One should never rate tools and equipment too high in his inventory, because if work suddenly stops they are almost worthless. Suppose a contractor has $10,000 worth of tools and equipment. If work stops, he has no use for them. Neither has anyone else because other contractors are also out of work. So the tools and equipment can't be used and can't be sold. One may store and maintain tools for a few years and then learn that they have become obsolete.

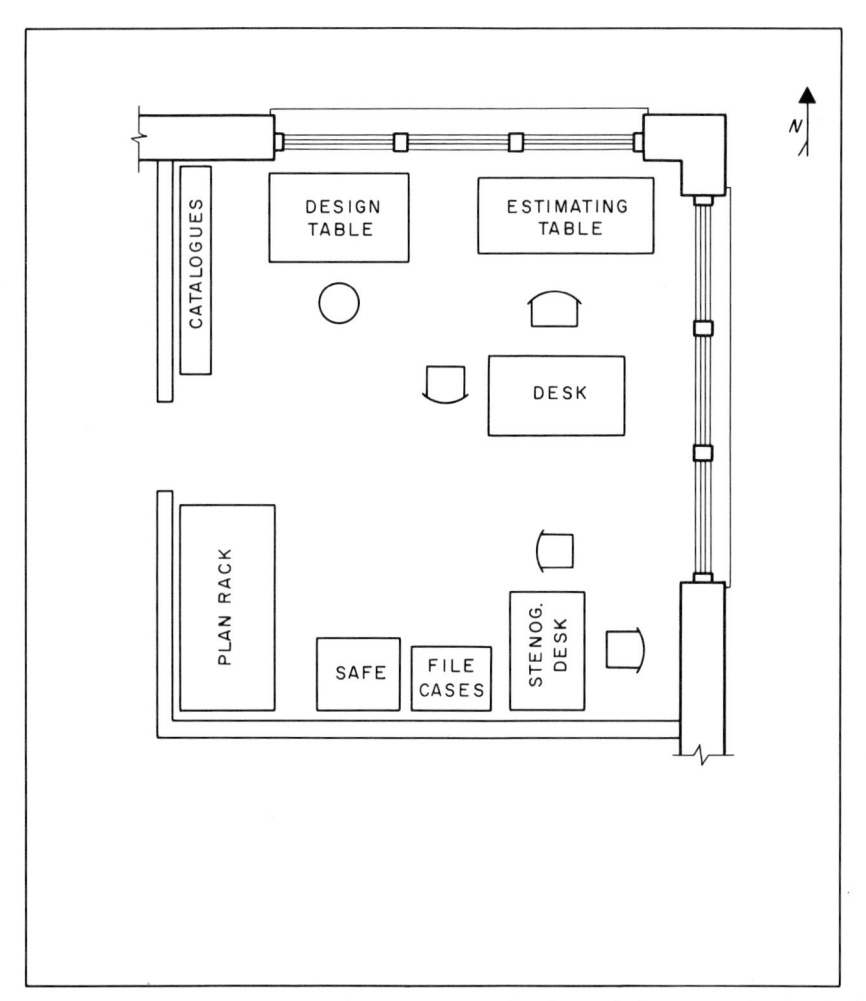

Fig. 1-1. Office requirements for a one-man business doing general electrical-construction work.

COMPETITION TO BE CONSIDERED

The competition with which one will be confronted must be considered. Too often, men going out for themselves fail to do this. They say, "There must be contractors to take the place of the old ones dropping out and I may as well be one of them." There are others who have the same idea, and many of them are men well trained in all phases of electrical construction.

It is true that from time to time the electrical contracting industry must have new blood. Fortunately, there is always a fair supply of well-trained men to fill this need.

As contractors grow old, they train men to take over their work. Numerous contractors all over the country are training members of their own families to carry on the business when they retire. Regularly, there are well-trained men who are outgrowing their existing positions and have no alternative except to go in business for themselves.

All beginners, good or bad, must buck the old-line competition. The old established firms have a smooth-running organization, ample tools and equipment of the best types, regular channels of business, and a reputation of long standing. These firms are out to get all the business they can and are not stepping aside for any newcomers.

Let it be stated again: The man contemplating a new venture cannot study the possible competition too carefully. Both the men who enter the electrical contracting business and those who decide to stay out benefit by such studies.

OTHER PHASES TO BE CONSIDERED

As we advance in future chapters, many things that must be studied by the man contemplating electrical contracting will be discussed. It will become obvious that the days when a mechanic can suddenly drop his tools and become a successful contractor are long since past. When the industry was in its infancy, the training required for a contractor was meager, tools few, competition limited, and the buying public tolerant. Today, competition, requirements of the industry, and demands of the buying public all place stringent demands on the small as well as the large contractor.

IT IS A GOOD BUSINESS

Electrical contracting is a good business. It has been deemed as such by Dun and Bradstreet and others. Many of those engaged in it consider it a very desirable vocation. That is why so many of the contractors are training members of their own families to take over the business. Once one has overcome the hazards of getting started and is established, he can operate with reasonable smoothness and impunity.

Contracting provides an opportunity for one to exercise his imagination, business acumen, and resourcefulness. It has its headaches the same as any other business, but when they are over, one enjoys the feeling of satisfaction that comes from being able to conquer obstacles.

Yes, electrical contracting is a good business. If you are properly trained, have the business connections and money to get started, and are willing to make the sacrifices, your chances of being a success are good. However, one must remember that essential aids to business acumen are confidence and enthusiasm.

CONFIDENCE AND ENTHUSIASM

To succeed, one must have confidence in what he is doing and go ahead wholeheartedly. We see new contractors as well as old who are trying to get along with practically no office equipment and inadequate tools. That is because they lack confidence. With lack of confidence goes lack of enthusiasm, and enthusiasm is a vital adjunct to any successful venture.

One cannot expect to gain the confidence of others when he is not sure of himself. His office and equipment are a criterion of his confidence in his undertaking.

CHAPTER 2

Do You Know How Much It Costs to Get Started in Business?

If you enter the electrical contracting business, do you know how much it will cost you to get set up and ready to operate? After you are ready to operate, do you know how much it will cost to keep your organization intact until you have enough money coming in from contracts to pay expenses?

If anyone were to tell you that to get started in a substantial line of electrical contracting your outlay would be around $18,000 to $20,000 before your operations could reach a steady paying basis, you would be astounded. You would probably dismiss the author of such a statement as a "crack-pot."

Many contractors have been able to get started for less than $18,000, but they were in a very limited line of work. Again others have started out to do a big job with a small amount of money, and they are still trying to get started.

You want to know what the $18,000 represents. We shall study a specific type of business and see what figures develop.

The cost of getting started in business depends on many things, the principal two being the type of business and the volume anticipated. Since all conditions cannot be covered, let us confine our attention to a one-man business, engaged in general electrical-construction work. By general electrical-construction work is meant limited-sized industrial and commercial projects and possibly some residential work.

What we mean by a one-man business can best be illustrated by listing an approximate division of the owner's time and explaining that he does all the estimating, engineering, selling, and administrative work. In fact, he would be the only one in the office besides the bookkeeper. The bookkeeper, in turn, would serve as a stenographer, answer the phone, and take care of the office in general.

Figure 2-1 shows the approximate division of the proprietor's time for a one-man business. There can be no formula for the division of a proprietor's time, as it must be applied to the various duties in ac-

A One-man Business
Approximate Division of the Proprietor's Time

Activity	Time per Week Percentage	Hours
Administrative	20	8.8
Estimating	25	11
Engineering and layout	10	4.4
Superintending	15	6.6
Selling and making contacts	15	6.6
Purchasing	5	2.2
Billing and collecting	5	2.2
Miscellaneous	5	2.2
Totals	100	44

Fig. 2-1. There is no formula for allotting a proprietor's time. The division of time for a business that is new varies greatly from that of an established firm.

cordance with the demands of the work in progress and the business in general.

ITEMS OF EXPENSE

The following represents a list of expenses to be met before the first draw on any work can be made:

1. Preparation expenses
   Office equipment and supplies
   Tools and construction equipment
   Truck (see text)
   Preparation of storeroom
   Stock
2. Operating expense while waiting for income from work
   Salaries (office and job payroll)
   Rent

Heat, light, and telephone
Cartage
Interest on investment
Insurances
Advertising and incidental expense

OFFICE EQUIPMENT

Figure 2-2 gives a listing of the minimum amount of office equipment required for the type of business we are discussing. Many items

FURNITURE REQUIRED FOR A CONTRACTOR'S OFFICE
Approximate Cost for a One-man Business

1 Owner's desk	$ 175
1 Stenographer's desk	150
1 File cabinet—three-drawer, 8½ by 11 in	80
1 File cabinet—three-drawer, 12 by 16 in	90
1 Table—estimating, approximately 72 by 42 in	60
1 Armchair—swivel type	60
1 Stenographer's chair—swivel type	35
4 Straight-back chairs	100
1 Design table—60 by 36 in	100
1 Plan rack—open pipe-frame type	50
1 Typewriter	170
1 Safe—small	450
2 Electric fans	50
1 Catalogue file—case type	50
1 Miscellaneous	100
Total	$1,720

FIG. 2-2.

ordinarily required are not included in the listing. These are such items as floor covering, water cooler, drapes, adding machine, drawing instruments, filing cabinet for tracings, and numerous others. Addition of these would greatly increase our estimated cost.

In addition to the office equipment there may be carpenter work and decorating to be taken care of. Almost without exception, one finds that when the bills are paid, the cost has run much higher than his original estimate.

One may be able to shop around and buy less expensive furniture. However, there is a limit to how far one can go. Shoddy furniture is poor advertising and indicates lack of confidence in the venture. As stated before, with lack of confidence goes lack of enthusiasm—a factor so vital to success in business.

TOOL COSTS

Figure 2-3 gives a list and estimated cost of tools required to equip a five-man crew working on an industrial project. Although the list is not exactly what a man starting out would expect to buy, it is a good estimate of what would be needed and gives an idea of minimum costs. Figure 2-4, a tool check list, shows many items normally in use but not included in the Fig. 2-3 listing.

Before one discounts the listing shown in Fig. 2-3, there are four things that must be taken into consideration, namely:

1. There are many tools commonly found on jobs that are not included in the listing.
2. To compete with other contractors, labor costs must be kept down by the use of modern and adequate tools.
3. The better mechanics are accustomed to working with good tools and expect to find them on the job.
4. The listing in Fig. 2-3 is for five men on one job. With the men divided on smaller jobs, many duplicate tools would be required.

It is true that one need not buy tools until he has a job to use them on, but regardless of when they are purchased or how they are paid for, they represent an outlay of money to get the business started.

TRUCK EXPENSE

In most locations where a contractor can keep four or five men regularly employed, there is some kind of trucking service available. Hence it is not always necessary to figure in the cost of a truck at the opening of a business. That is true for the type of electrical-construction work being discussed. However, there are types of work, such as the installation and maintenance of highway signals and general maintenance work, that require a truck.

PREPARATION OF STOREROOM

For the purpose of estimating, we shall assume that the contractor buys material as needed for each project and has it delivered directly to the job. Thereby he will be able to keep down the outlay for stock and the cost of equipping the storeroom. Regardless, there would have

ESTIMATED COST OF TOOLS TO EQUIP AN INDUSTRIAL JOB
REQUIRING A CREW OF FIVE ELECTRICIANS

Item	No. in Use	Purchase Price Each	Purchase Price the Lot
Electric drills—¼ in.	1	$ 45.00	$ 45.00
Electric drills—½ in.	1	70.00	70.00
Stepladders—8 ft.	1	15.00	15.00
Stepladders—10 ft.	1	20.00	20.00
Stepladders—12 ft.	2	24.00	48.00
Stepladders—14 ft.	1	35.00	35.00
Pipe benches—small	2	23.00	46.00
Pipe benches—large	1	30.00	30.00
Pipe vise—½ to 2 in.	2	11.00	22.00
Pipe vise—2 to 3½ in.	1	19.00	19.00
Stocks—½ to 1 in.	2	17.00	34.00
Stocks 1¼ to 1½ in.	1	25.00	25.00
Stocks—2-in. ratchet	1	38.00	38.00
Stocks—2½ to 4 in.	1	115.00	115.00
Pipe dies—½ to ¾ in.	4	3.60	14.40
Pipe dies—1 to 1¼ in.	2	5.50	11.00
Pipe dies—2 in.	1	6.00	6.00
Pipe dies—2½ in.	1	10.00	10.00
Pipe dies—3 in.	1	10.00	10.00
Pipe dies—4 in.	1	12.00	12.00
Hickies—miscellaneous ½ to 1 in.	5	5.00	25.00
Pipe benders—small	1	200.00	200.00
Toolboxes—large wood	1	50.00	50.00
Toolboxes—steel	2	35.00	70.00
Rope—hemp ½ in., 100 ft.	1	4.00	4.00
Rope—¾ in., 100 ft.	1	7.50	7.50
Scaffolding—rolling	1	250.00	250.00
Wagon trucks	2	50.00	100.00
Chain hoist—5 ton, 10 ft lift	1	270.00	270.00
Extension cords—50 ft, heavy duty	2	8.00	16.00
Gas furnace—plumbers	1	35.00	35.00
Pipe wrenches—18 in.	2	5.00	10.00
Pipe wrenches—24 in.	2	8.75	17.50
Chain tongs	2	18.00	36.00
Reamers—miscellaneous ratchet	3	15.00	45.00
Oilers	4	0.50	2.00
Knockout punches	1	10.00	10.00
Locks and chains	4	4.00	16.00
Hacksaw blades	30	0.10	3.00
Files—miscellaneous	5	0.70	3.50
Hammers, drills, and miscellaneous		20.00	20.00
Total			$1,815.90

FIG. 2-3.

TOOL LIST
FOR
ELECTRICAL CONSTRUCTION (INTERIOR)

JOB_____ JOB NO._____
ADDRESS_____

	NO.	SHIPPED	RETURNED		NO.	SHIPPED	RETURNED		NO.	SHIPPED	RETURNED
I.- PIPE TOOLS				**III.- GEN. CONST. TOOLS**				**IV.- WIRE INSTALLATION & CONN.**			
BENCHES LARGE				TOOL BOXES - STEEL							
BENCHES SMALL				TOOL BOXES - WOOD				FISH TAPES 1/8 IN.			
VISES 1/2" TO 2"				LOCKS				FISH TAPES 3/16 IN.			
VISES 2" TO 4"				CHAINS - LADDER				FISH TAPES 1/4 IN.			
STOCKS				LADDERS - STEP 6 FT.							
1/2" TO 1"				LADDERS - STEP 8 FT.				WINCHES - HAND DR.			
1 1/4" TO 1 1/2"				LADDERS - STEP 10 FT.				WINCHES - POWER DR.			
2" RACHET				LADDERS - STEP 12 FT.				POWER DR. - UNIVERSAL			
2 1/2" TO 4"				LADDERS - STEP 14 FT.				WIRE ROPE 1/4 IN.			
				LADDERS - STEP 16 FT.				WIRE ROPE 3/8 IN.			
DIES 1/2"								CABLE PULLERS "COME ALONG"			
DIES 3/4"				LADDERS - EXTEN.							
DIES 1"				SCAFFOLDING MAT.				REEL JACKS			
DIES 1 1/4"								PICK UP CART			
DIES 1 1/2"				EMERY WHEELS				BRAKE CABLE			
DIES 2"				ELECT. DRILLS 1/4 IN.				REELS - PAY-OUT			
DIES 2 1/2"				ELECT. DRILLS 1/2 IN.				REELS & STAND - GANG TYPE			
DIES 3"				DRILLS - TWIST SIZE							
DIES 4"				DRILLS - TWIST SIZE				STRIPPERS - WIRE			
THREAD CUTTER - PR. DR.				TAPS SIZE				GAS FURNACE			
BENDER SMALL				TAPS SIZE				SOLDER POT & LADLE			
BENDER LARGE				WHITNEY PUNCHES				ELECT. SOLDERING SET			
HICKIES (H.W. COND.) 1/2"				K.O. PUNCHES 1/2" TO 1"							
HICKIES (H.W. COND.) 3/4"				K.O. PUNCHES 1 1/2" TO 2"				STEEL LETTERING SET			
HICKIES (H.W. COND.) 1"				AIR HAMMERS							
THIN WALL BENDER 1/2"				ELECT. HAMMERS - NO.							
THIN WALL BENDER 3/4"				ELECT. HAMMERS - NO.							
THIN WALL BENDER 1"								**VI.- PAINTING TOOLS**			
				HACK SAW BLADES				WIRE BRUSHES			
WRENCHES 18"				REAMERS - TAPER				PAINT BRUSHES IN.			
WRENCHES 24"				REAMERS - TAPER				PAINT BRUSHES IN.			
CHAIN TONGS 1/2" TO 2 1/2"				FILES KIND & SIZE				BUCKETS - PAINT			
CHAIN TONGS 2 1/2" TO 4"				FILES KIND & SIZE				STENCIL PLATES			
REAMERS 1/2" TO 1 1/4"				STAR DRILLS SIZE							
REAMERS 1/2" TO 3"				STAR DRILLS SIZE				**VII.- TEL. & SIGNAL EQUIP**			
OILERS				BULL POINTS SIZE							
				BULL POINTS SIZE							
				COLD CHISELS							
				CROW BARS							
				PINCH BARS							
				"C" CLAMPS				**VIII.- MISCL. TOOLS & EQUIP.**			
II.- SHOP TOOLS (FIELD SHOP)				SOCKET WRENCH IN.				PICKS			
DRILL PRESS				SOCKET WRENCH IN.				SHOVELS			
POWER SAW				OILERS				WHEELBARROWS			
COPPER BENDER				EXTEN. CORDS				HOSE			
IRON BENDER (UNIVERSAL)				BOLT STOCK & DIE				BUCKETS - WATER			
EMERY WHEEL (ELECT.)								TARPAULINS			
WORK BENCH				BURNERS - GAS & TANKS				RUBBER GLOVES			
MACHINIST VISE				COFFING HOIST				RUBBER BOOTS			
				CHAIN HOIST				RAIN COATS			
ELECT. WELDER				ROPE - HEMP 1/2 IN.				GOGGLES - WELDING			
FORGE				ROPE - HEMP 3/4 IN.							
ANVIL				BLOCKS							
"C" CLAMPS											
				WAGON TRUCKS				**IX.- TESTING EQUIPMENT**			
				DOLLIES							
				JACKS							
				SLEDGES							
				CHAIN - LOG				**X.- MISCL. SUPPLIES - CHECK LIST**			
V. PORTABLE TOOLS - SPECIAL											
FLOOR CRANE								SOLDER PAINT			
AIR COMPRESSOR				SAWS - TWO MAN				TAPE VARNISH			
WELDER - ELECT.				SAWS - HAND				PASTE - SOLDER EMERY CLOTH			
FLEX. SHAFT EMERY - EL. DR.				SAWS - SKILL				OIL WASTE			
ROLLING SCAFFOLD				WOOD BITS				GASOLINE CHALK - CARPENTERS			
				WOOD BITS				GRAPHITE CRAYON - CARPENTERS			
				EXTEN. BITS				METAL TAGS			
				WOOD CHISELS							
				SQUARE - CARPENTERS							
				LEVEL - "							
				AXE							
				CHALK LINE							

Fig. 2-4.

Do You Know How Much It Costs to Get Started in Business? 15

to be bins for small fittings, fuses, small tools, etc. There would also have to be a workbench, shop tools, and wire reels. There would have to be special lighting for the bins and benches.

For our estimate we shall allow $500 for the storeroom preparation. This is a very small amount compared with the amount spent

FIG. 2-5. Rolling scaffold and platform hoist.

for storerooms and shops equipped with lift trucks, wagon trucks, portable hoists, numerous small tools, and racks and bins for the storage of large quantities of stock.

STOCK

To take advantage of quantity discounts and package prices, the contractor must carry some stock. The stock may not be paid for immediately, but it must be figured in with the starting cost.

FIG. 2-6. Band saw for cutting metal.

FIG. 2-7. Hydraulic pipe bender.

Fig. 2-8. Platform hoist.

Fig. 2-9. Semitrailer serves as a portable stockroom.

Fig. 2-10. Truck with hydraulic platform lift.

Fig. 2-11. Panel truck.

Often contractors start by buying a fair amount of stock, but we are figuring a minimum cost for getting started, so shall make a small allowance. For a starting price let us use $300. A supply catalogue or material check list will show that there are numerous small items that must be stocked. In addition to small fittings one must have some conduit, wire, and trim fittings in stock. He cannot expect a supply house to make a trip to a job that is finishing up with a few small items that have been overlooked.

Fig. 2-12. Special truck with overhead pipe and ladder rack.

Before one is operating on a paying basis, there will be stock coming back from completed projects. This cannot be avoided because job requirements vary. The stock returned not only will represent a small investment but will require storage facilities. Materials not properly stored cannot be readily located.

OFFICE LOCATION AND RENTS

Contractors are learning that having an office in the heart of a city is not important. The principal advantage of being centrally located is the easy access to the offices of architects and engineers. A location in an outlying district offers many advantages to offset this one.

Not many years back there were no banks in outlying districts and

MATERIAL CHECK LIST

CONDUIT	COVERS - (OUTLET BOX)	METERING EQUIPMENT	SIGNAL & CLOCK EQUIP.	
CONDUIT - HEAVY WALL	PLASTER RINGS	METER FTGS.	TRANS. - BELL	CHIMES
CONDUIT - THIN WALL	SWITCH COVERS - (4" SQ.)	METER BOARDS	CAB - TRANS.	ANNUNCIATORS
FIBER DUCT	GANG SW. COVERS	CURRENT TRANS.	PUSH B. - WALL	TEL.
COMPOSITION CONDUIT	BRACKET COVERS	WATT - HOUR METERS	PUSH B. - DESK	
	4" SPIDER SW. COVERS			MAIL BOX & P B
	BUSHED COVERS		BATTERY	VESTIBULE TEL.
ELBOWS & CPLGS - H.W. CONDUIT	BLANK COVERS	**SWITCH BRDS. & PANELS**	CHARGER	DOOR OPENER
" " - T.W. CONDUIT	4-11/16" SQ. COVERS - (BLANK & MISC.)		CHARGER PANEL	TIME STAMPS
" " - FIBER DUCT	FITTING COVERS	MAIN SW. BOARDS	CABINETS - TERM.	SIGNAL RELAYS
" " - COMP. CONDUIT		HEADER BOX - SW. BOARD	TERM. STRIPS	PAGING INST.
		BASE CHANNEL - SW. BOARD	CORD - (DESK P B)	FIRE ALARM
	MISC. ROUGHING MATERIAL	GRILL & DOORS - SW. BOARD		
OUTLET BOXES		RUBBER MATS - SW. BOARD	BELLS	CLOCKS EL.
	BAR HANGERS		BUZZERS	TIME CLOCKS
4" OCT. ------(8B)	FIXTURE STUDS	DIST. PAN. & C. - LTG & PR.		
4" OCT. ---DEEP--(8B DEEP)	BOX ANCHORS	BR.CT. PAN. & C. - LTG.		
4" SQUARE ---- (1900)	COMB. COND. ANCHOR & FIXT. STUD	BR.CT. PAN. & C. - PR.	**MISC. MATERIAL**	
4" EXTENSION RINGS	SPCL. FIXTURE SUPPORTS			
4-11/16" SQ. BOXES -- (11B)	TILE ARCH OUTLET SUPPORTS	EMER. THROW OVER	UNDER FLOOR DUCT	
CONCRETE BOXES		AUTOMATIC SWITCHES	AUTO-TRANS.	
CAST IRON BOXES	METAL MLDG. & FTGS.	METERING PANELS	CABLE PIT FRAMES	
FLOOR BOXES	OVAL DUCT & FTGS.	FIRE P. PANELS	CABLE PIT COVERS	
DOOR SW. BOXES		TRANS. PAN - (METERING)	CABLE PIT INS. SUPPORTS	
GANG SW. BOXES			FIRE - PROOFING MAT.	
SECTIONAL SW. BOXES	GREENFIELD	SERV. SWS.	CONC. MAT. - (U.G. DUCT)	
HANDY BOXES	GREENFIELD FTGS.		DUCT SPACERS	
FITTINGS - THREADED	B - X	FUSES - SERV, FEEDERS,& BR. CTS.	CABLE SPLICING MAT.	
FITTINGS - NO THREADS	B - X FTGS.	N.E.C - ONE TIME PLUG	BOLTS, NUTS, MACHINE SCREWS	
		N.E.C - RENEWABLE SPECIAL	WIRE - (IRON)	
EXPLOSION - PROOF BOXES				
	WIRE & CABLE			
PULL BOXES & TUBS	R.C. --- 600 V.	**POWER CONN. EQUIP.**	**SPECIAL LISTING**	
	L & R.C. --- 600 V.	SWITCHES	TO BE FILLED IN BY INDIVIDUAL CONTRACTOR	
PULL & HEADER BOXES	V.C. --- 600 V.	STARTERS		
TERMINAL CAB.	V.C. - L.C. --- 600V.	PUSH BUTTON		
TRANS. CAB. - (METERING)	SERV. ENT. CABLE			
TRANS. CAB. - (SIGNAL)	HIGH VOLTAGE CABLE	STARTER FRAMES		
CABLE SUPPORT BOXES		FLEX. CONDT.		
SPCL. SERVICE HEADS	BELL WIRE	FLEX. CONDT. FTGS.		
WIRING TROUGHS	TELEPHONE CABLE	GROUNDING MAT.		
TUBS FOR FLUSH FIXTURES	TELE. - TWISTED PR.	PADS - CORK		
SPCL. SW. BOXES		FOUND. MAT.		
		TROLLEY DUCT.		
	FIN. MAT. - LTG. BR. W.	PR. REC. & PLUGS		
FITTINGS	S.P. SWITCHES	PR. CABLE - (SUPER - SERVICE)		
PIPE SLEEVES	3W & 4W SWITCHES	BUS DUCT PLUG - INS		
LOCK - NUTS & BUSHINGS	LOCK SWITCHES	FUSES		
INSULATION BUSHINGS	CEILING PULL SWITCHES	CRANE TROLLEY EQUIP.		
COND. SUPPORTS - (RISER FL. CLIPS)	PENDANT SWITCHES			
HANGER MAT. - (TRAPEZE)	SPCL. SW. - EXPL. PR., ETC.	**FIXTURES & LAMPS**		
HANGER RINGS & RODS				
PERFORATED IRON	RECEPT. - DUPLEX	RLM UNITS	COVER SOCKETS	
STRAPS - PIPE	RECEPT. - SINGLE	MISC. STEEL FIXT.	DROP CORDS (COMPL.)	
MIN. CLAMPS	RECEPT. - PILOT LP	GLASS STEEL UNITS	LAMP GUARDS	
TOGGLE BOLTS	RECEPT. - SPCL.	FLOURESCENT	FIXT. ALIGNERS	
SCREWS - WOOD - LAG		SLIM LINE	HICKIES & STUDS	
NAILS	FAN HGR. OUTLET	COLD CATHODE	CANOPY SW.	
	CLK. OUTLET	MERCURY VAPOR	WINDOW LTG. EQ.	
	COVER SOCKET	MISC. OFFICE UNITS	CASE LTG. EQ.	
BOX CONN. - THIN WALL CONDT.	RANGE RECEPT.	BRACKET	EXIT LTG.	
COUPLINGS - " " "	W.P. RECEPT.	VAPOR PROOF	FIXTURE WIRE	
SERVICE HEADS	EXPL. PR. RECEPT.	EXPLOSION PROOF	FIXTURE STEMS	
FITTINGS & COVERS - CONDT.	PLATES - SW., REC., FTG., ETC.	FLUSH LTG. UNITS	TAPE, SOLDER, ETC.	
END BELLS		EXTERIOR LTG.	WIRE NUTS	
ADAPTERS - CONDT. TO FIBER DUCT	REC. CAPS	FLOOD LTS.	LAMPS	
ADAPTERS - H.W. TO THIN WALL	WIRE NUTS			
PIPE CAPS	BUSHINGS - ENT. CAB.			
SLEEVES				
NIPPLES - COND. - CHASE	**MISC. FEEDER & GROUNDING**			
REDUCERS				
INSERTS & EXP. SHELLS	BUS DUCTS			
CINCH ANCHORS	CABLE SUPPORTS			
BEAM CLAMPS	GROUNDING MAT.			
	SOLDERLESS CONN.			

FIG. 2-13. Material check list.

industries were concentrated in the cities. Under such conditions there were many reasons for having an office in the city proper. Today industries are widely spread out and outlying banks are common.

Let us look at some of the advantages of having an office in an outlying district:

1. Rents are lower.
2. It is easier to find a location where the shop and stockroom can be located adjacent to the office.
3. Travel time between home and office can be greatly reduced.
4. There is seldom need for driving into the city before the rush-hour traffic has abated.
5. When business falls off, one is not caught with a lease for high-rent space on his hands.

To this list, others of lesser importance could be added.

The one item in the above listing that justifies elaboration is that of having the stockroom adjacent to the office. For a one-man business, this is imperative. Either the owner or the office man must look after the stock and out-of-stock shipments.

There are other advantages in having the shop close to the office; however, since we are principally interested in costs, they will not be discussed here.

So far we have pretty well covered the preparation costs, which can be summed up as follows:

1. Office equipment and supplies	$1,700
2. Tools and construction equipment	1,800
3. Preparation of storeroom	500
4. Stock	300
Total	$4,300

Here we have a total of $4,300. To this list might be added carpenter work, decorating, floor covering, air conditioner, water cooler, and many other items of cost. The total could easily be built up to $5,000 or $6,000. Besides, nothing has been included for the owner's salary during the preparation period.

We have been discussing just one division of cost. The owner must also advance money for operating costs while waiting for contracts.

CHAPTER 3

*Do You Know How Much It Will
Cost You per Month While You
Are Waiting for Contracts?*

Let us not go into this blindly as so many do. We shall make up an estimate of costs and include a minimum allowance for the proprietor's salary.

Many men do not include any salaries for themselves. They say their salaries will come out of the profits. How absurd. Have you ever found any of these men who, for the final markup, put on the label For Salary and Profit? Money used for salary is not left for profit, and if it is profit, it cannot be used for salary.

Ignoring an expense will not cancel it out. Too many entering business delude themselves and try to keep their courage up by ignoring costs that cannot be escaped. That is a critical weakness and not healthful for good business.

We shall include a salary for the owner. The question is, How much? We are writing a book which, if it experiences a circulation similar to its companion book *Electrical Estimating,* will reach readers in all parts of the United States as well as in many foreign countries. And all these readers are in communities that have different wage rates and operating costs.

Salaries allotted proprietors are usually in proportion to the wage rates of the local electricians. It is hard to get information on small communities; however, it seems safe to say that wages for open-shop electricians run as low as $1.50 per hour. Contrasted to this, we have union rates in large cities up to $4 and more. For example, effective

July 1, 1960, union rates were as follows: Montreal, Canada, $2.45; Winnipeg, Canada, $2.75; Minneapolis, Minn., $3.70; Chicago, Ill., $4.30.

The reader can understand why one must be concerned about base wage rates used for figuring operating costs. If low rates are used, the reader from the large city may consider the estimated costs absurd. If high rates are used, the man in the low-cost community will think

Estimated Operating Cost (Minimum) per Month for Operating a One-man Business	
Proprietor's salary	$ 700 *
Bookkeeper and general office man	350 *
Rent (heat included)	150
Light	10
Telephone	15
Depreciation on equipment (not charged to jobs)	15
Interest on investment	15
Automobile and travel expense	85
Insurance and miscellaneous	15
Total estimated cost per month	$1,355
Estimated cost per year	$16,260

* Salary would be low for a high-wage-rate community.

Fig. 3-1.

that someone is trying to frighten him out of going into the electrical contracting business.

For our estimates, $3 per hour will be used as the electrician's rate and the proprietor's salary will be in some proportion. The reader can use the estimates as a guide and substitute wage rates suitable to his community.

Normally it is assumed that the proprietor should allow himself a salary 50 to 100 per cent greater than that paid electricians. Since we are figuring rock-bottom prices, only one-third will be added and $4 per hour will be used as the owner's rate of pay.

A 40-hr week represents approximately 173 hr per month. At $4 per hour, the salary would be $692 per month; $700 will be used without stipulating any definite hours. Most electrical contractors starting out work much more than 40 hr per week.

Figure 3-1 lists the principal operating costs that go on regardless of whether there are any contracts or not. The total amount given is $1,355. When business is operating at capacity, some of the items of expense will be increased.

How Much Will It Cost You per Month While Waiting for Contracts? 25

At the beginning, there will be no income to absorb the operating costs. It will not be until draws are made on contracts that the contractor can cease advancing the money for them.

In addition to the operating expenses which continue to pile up, there will be money advanced for payrolls and some material costs. We shall assume that the contractor waits a reasonable length of time for job collections before he pays for the material. He will more than likely lose his discounts, but that will not be recognized here.

There may be some accounts that are very slow in paying bills. In such cases the contractor will be obliged to pay the supply house before he makes collections. On overdue bills, both material and payroll costs will be tied up.

THE FIRST SIX MONTHS

To get something to work with we can make an estimate of the first 6 months' business. The resulting figures can be used in our calculations.

Figure 3-2 shows an estimate for the first 6 months in business. It will be noted that of the combined cost of material and labor, 60%

ESTIMATED FIRST SIX-MONTH CONTRACTS *
FOR A ONE-MAN BUSINESS

		Material		Labor
First month—no business		—		—
Second month—average one man employed		$ 780		$ 520
Third month—average two men employed		1,560		1,040
Fourth, fifth, and sixth months— average four men employed		9,360		6,240
Totals		$11,700		$ 7,800
Direct job expenses (not including insurance)	2%	234	6%	468
Job costs total		$11,934		$ 8,268
Overhead	10%	1,193	35%	2,894
Cost including all operating expenses		$13,127		$11,162
Return	5%	656	10%	1,116
Estimated sell material		$13,783		$12,278
Estimated sell labor		12,278		
Total estimated contracts		$26,061		

* Based on a wage rate of $3 per hour for mechanics. At $4 per hour the amount would be 33⅓ greater.

FIG. 3-2.

has been allowed for material and 40% for labor. It must also be noted that very substantial markups have been used.

If one starting out were confined to work obtained from competitive bidding, it is very doubtful if he could get much work with markups in excess of those required to cover bare costs. The reader must again be reminded that a beginner is not likely to be able to carry on his work so efficiently as the going concerns. This is another handicap to bidding.

Suppose one were fortunate enough to get the amount of work, with the liberal markups, shown in Fig. 3-2. He would have direct job sales of material $11,934 and labor $8,268, making a total of $20,202. The estimated sell price is $26,061. Subtracting the direct job sales from the sell price leaves $5,859 to cover all operating expenses and return.

The operating cost per year (see Fig. 3-1) is $16,269. For 6 months they would be one-half that amount, or $8,130. Subtracting $5,859 (total markup on contracts) leaves $2,271. The contractor has spent approximately $2,271 more than his contracted income. When we are figuring his total outlay, $2,000 will be used because the markups used for job costs included allowance for engineering and estimating, which will be included in the proprietor's salary.

THE FIRST YEAR

In estimating the total outlay of money for the first year, it will be assumed that the operating-expense losses for the second 6 months will be half of that for the first 6. As $2,000 was established for the first 6 months' losses, $1,000 will be used for the second.

An estimate of the total outlay, in excess of income, for the first year, is as follows:

1. Preparation expenses:		
Furniture and office equipment............	$1,700	
Tools and construction equipment..........	1,800	
Storerooms and stock.....................	800	
Total preparation cost..................		$ 4,300
2. Operating expense—excess:		
First 6 months...........................	$2,000	
Second 6 months..........................	1,000	
Total operating expense losses...........		$ 3,000
3. Cost of work not collected for..............		2,000
4. Payroll—current jobs (5-man mo. not billed)..		2,600
Total................................		$11,900

Previously it was stated that a normal cost of getting started would be $18,000 to $20,000. Here we have $11,900 for the first year. The difference between $11,900 and $20,000 is accounted for as follows:

1. It is often 2 or 3 years before contractors get enough work to put the business on a paying basis.
2. The markups used in the estimates were higher than those normally enjoyed by contractors new in the business.
3. The allowance for contracts secured is liberal.

Fig. 3-3. In planning an office, one should allow for future expansion and additional equipment to be acquired when the business can afford it.

It is awfully hard for one contemplating a new business to realize how rapidly time will slip by while he is waiting for contracts. He will go to the office of a friendly architect or engineer and be told that in a couple of months they will have plans out for figures. The couple of months pass, and it is 4 months before the plans are ready for figures. The work is figured, and it is another month before any contracts are let. Under normal conditions, for competitive bidding, chances of getting the contract for electrical work are only 1 out of 5.

If one is not careful and fortunate, a year can slip by and he will find that he has had very little remunerative work. In the next chapter a study will be made of the volume of contracts needed to support the type of one-man business being studied.

CHAPTER 4

Do You Know How Much Business Must Be Forthcoming to Pay Operating Expenses?

Do you know that a one-man business engaged in general electrical-construction work would need a $100,000 volume in order to be able to pay all operating costs and provide for a small profit? Furthermore, this volume would have to be made up of contracts having liberal markups.

In a previous chapter an example (Fig. 3-1) provided an estimate of operating expense (not including job tools) per month and per year for a new business. Figure 4-1 gives the operating costs per year for an established business. The estimate lists all expenses including consumed and depreciated tools.

Figure 4-1 does not segregate the direct job costs from the items of overhead expense. To do this would increase the listings and promote confusion. For example, the proprietor's salary would have to be divided between executive duties and direct job costs such as superintending, engineering, and layout work.

Later, a study of operating costs will be made. Here let it suffice to say that:

1. In general, operating costs are represented in two divisions, namely: (*a*) direct job costs, (*b*) overhead.
2. Any expense that can be identified with a particular job is a direct job cost.

3. Expenses which cannot be identified with any particular job are classed as overhead.

We know from previous studies that it takes approximately $100,000 (sell) to justify $18,000 operating expense. We know also that the base cost of such a volume with an M/L ratio of 60/40, would be $45,000 (60%) for purchased material and $30,000 (40%) for payroll. Figure 4-2 provides an estimate for the total volume based on these costs.

A ONE-MAN BUSINESS

*Estimated Operating Costs for an Established Business
Based on a $100,000 Volume (Annual)*

Proprietor's salary	$ 8,400 *
Bookkeeper and general office man	4,200 *
Rents—office and storeroom (heat included)	1,800
Light	120
Telephone	240
Automobile and travel expense	1,000
Stationery and supplies (office)	125
Interest on investment	200
Depreciation on equipment (not including tools)	200
Advertising and legal expenses	300
Postage and miscellaneous	250
Association dues and licenses (direct job expense)	500
Tools (a direct job expense)	750
Total	$17,885

* These are estimating allowances. One would expect to find salaries much higher in large United States cities.

Fig. 4-1. Operating costs vary greatly, depending on the type of work contracted for.

The summary of costs, Fig. 4-2, lists the total, with all operating expenses figured in, as being $18,420. This is $535 more than the estimated operating costs shown in Fig. 4-1. However, the two figures are close enough together to verify the statement that a $100,000 volume will be needed.

If one wanted to arrive at figures that would coincide, it could be easily accomplished. Many of the items listed in Fig. 4-1 could be increased and several items of expense could be added, such as:

1. Cost of carrying men on the payroll between jobs
2. Bad debts
3. Stolen tools and equipment

4. Adjustment factor (increased overhead to compensate for losses during bad times)

The estimate, Fig. 4-1, includes $8,400 for the owner's salary. A man capable of running a general electrical-construction business should allow himself twice that amount. He could earn much more than $8,400 working for someone else. It must be remembered that

---

A ONE-MAN BUSINESS

Approximate Distribution of Costs for a $100,000 Volume (Annual)

	Material		Labor
Purchase price—material and payroll.........	$ 45,000	60%	$30,000 (40%)
Direct job costs (insurance not included) (2%).	900	6%	1,800
Total job costs..........................	$ 45,900		$31,800
Overhead........................ (10%)	4,590	35%	11,130
Total cost including overhead.............	$ 50,490		$42,930
Return (anticipated profit)............ (5%)	2,525	10%	4,293
Sell price material...................	$ 53,015		$47,223
Sell price labor.....................	47,223		
Sell price the lot....................	$100,238		

Included for operating costs:
- Direct job cost—material........ $ 900
- Direct job cost—labor..........   1,800
- Overhead—material............   4,590
- Overhead—labor..............   11,130
- Total...................... $18,420

FIG. 4-2.

---

we used the low allowance for salary because minimum amounts were being included.

One not familiar with the separate markup method of estimating may wonder why overhead markups of 10% for material and 35% for labor were used. Details will be explained later, but briefly, on a dollar basis, the cost of supplying labor is much greater than the cost of supplying material. A common markup of 20% on a 60/40 cost (60% material and 40% labor) produces the same results as separate markups of 10% on material and 35% on labor. Common markups on costs with M/L ratios other than 60/40 would not produce true costs.

Before we leave the subject of volume required, another example must be studied to emphasize the effects of higher operating expense and low markup. Figure 4-1 showed an operating cost of $17,885 per year. For the new example, the owner's salary will be increased by $5,000 and the allowance for other expenses will be increased approximately 10 per cent. The revised estimate of operating costs will be as follows:

$$
\begin{aligned}
&\text{Original estimate} \ldots \ldots \ldots \ldots \ldots \ldots \$17,885 \\
&\text{Salary adjustment} \ldots \ldots \ldots \ldots \ldots \ldots 5,000 \\
&\text{Other increases} \ldots \ldots \ldots \ldots \ldots \ldots 950 \\
&\text{Total estimated operating cost} \ldots \$23,835
\end{aligned}
$$

The revised estimate of operating expense is $23,835. To cover this expense in our new estimate of volume, the much talked about 10% for markup will be included. Many poorly informed contractors use 10% and 10% (10% for operating costs and 10% for profit) markups. How such contractors are able to stay in business will be discussed later.

With operating costs amounting to $23,835 and an allowance of 10% markup to cover them, an estimate of the year's volume of business would be:

$$
\begin{aligned}
&\text{Material} \ldots \ldots \ldots \ldots \ldots \ldots \ldots \$143,010 \ (60\%) \\
&\text{Labor—payroll} \ldots \ldots \ldots \ldots \ldots 95,340 \ (40\%) \\
&\phantom{\text{Labor—payroll} \ldots \ldots \ldots \ldots \ldots} \$238,350 \\
&\text{Markup for operating expense } 10\% \ldots 23,835 \\
&\text{Total} \ldots \ldots \ldots \ldots \ldots \ldots \ldots \ldots \$262,185
\end{aligned}
$$

The $262,185 volume is more than double our original estimate, and there is nothing included for profit. It would have to be an unusual situation for a one-man business to take care of such a volume. It is only with large-volume contracts that low markups can be used.

When operating costs and markups are being studied, it will become evident that one cannot operate his business by hearsay. He must learn his own costs and operate his business according to his own experience.

CHAPTER 5

Do You Know Where You Are Going to Get Contracts for $100,000?

Before starting the main text of this chapter, let it be stated again that the author is not attempting to discourage anyone contemplating electrical contracting. The getting of contracts is a major problem, and it is well to look it square in the face before opening an office. It is a problem that is always with contractors regardless of the length of time they are in business.

Do you know where you are going to get contracts for $100,000 worth of business? A $100,000 does not impress one as being much when he thinks of the millions that are being spent for electrical construction work, but when he starts from scratch, the amount will soon begin to loom up as a considerable sum.

It is so easy for some men, particularly superintendents and mechanics, to get the idea that most contracts come automatically. They have been working for the same contractor for a long time, and when one job is completed, they are told to go to another. This goes on year in and year out. Is it any wonder that they get the idea that there is always plenty of work?

The complaints of contractors about wild bidding and cutthroat competition do not influence the opinions of most of us. Complaining about competition has been a regular stock in the trade of electrical contracting as long as we can remember. Contractors complain consistently but stay in business and make money. It is an "awful business," yet they are anxious to have members of their families get into it. Is it any wonder that mechanics are not impressed by their complaints?

Seasoned electrical estimators are not so likely to overlook the problem of getting contracts. However, often some of them fail to understand how much the contractor they are working for is dependent on established connections. It seems reasonable to say that 50% of the work done by electrical contractors is obtained, directly or indirectly, through old established business connections.

BUSINESS MUST BE ASSURED

A man engaged in the type of business we are studying should be assured of approximately $50,000 in volume, without competition. To reach the $100,000 volume, he would still have to obtain another $50,000 worth of business on the competitive market.

Estimating time would be a limiting factor for contracts earned in competitive bidding. In the division of the proprietor's time as shown in Fig. 1-1, 11 hr per week was allotted to estimating. Figuring 50 work weeks per year and 11 hr per week would allow 550 hr per year for estimating. This would be almost entirely consumed in figuring enough competitive work to get $50,000 in contracts. The figures are as follows:

Under normal conditions a contractor is able to secure about one out of five jobs figured. Based on one out of five, he would have to estimate $250,000 worth of work to obtain $50,000 in contracts. These contracts would more than likely be in the $1,000 to $10,000 bracket. Two hours per $1,000 is normal estimating time.

To figure $250,000 worth of work at 2 hr per $1,000 would require 500 hr. Only 550 hr has been allowed for estimating all work, competitive and otherwise.

DO YOU HAVE BUSINESS CONNECTIONS?

Again, do you know where you are going to get $100,000 worth of work? Do you have friends that can give you work, and do you know architects and engineers who will be glad to have you figure their work? If you get jobs to figure, are you reasonably sure that you have an even chance with your competitors of being considered for the contract? All these questions must be weighed before one opens an office.

One cannot bank too much on promises. He may have a friend in an industrial plant who promises him work. The friend may be permitted to let contracts to old established firms, but when a new contractor is considered, his firm may object. Again, by the time the contractor is organized and ready to go, his friend may have been shifted to another department or away from the plant altogether.

Architects and engineers may permit a contractor's name to be on their bidding lists long before they are confident that he is sufficiently organized to have a contract. Considerable time may elapse before they have a contract that they feel the contractor is equal to.

We do not like to think that there are men without scruples in the construction business, but we have plenty of reasons to believe that general contractors often have electrical contractors figure work when they have no intention of giving them the contract for electrical work. The general hopes to get a low bid which can be used to beat down the price of some established electrical contractor.

Before one gets excited about there being plenty of electrical work because millions are being spent for new construction, he must recognize a few facts:

1. Only a small part of the work will be electrical.
2. He will be eligible to figure only a small part of the electrical work.
3. There will always be keen competition because the established firms are trying to get all the business they can.
4. Established contractors may be able to carry on the work more expeditiously.
5. The established firms may get more consideration from buyers.

MANY CONTRACTS NEEDED

In terms of small jobs, $100,000 is a lot of work. Suppose one were to get contracts for $100,000 in the following amounts:

	3—average	$10,000	$ 30,000
	5—average	5,000	25,000
	45—average	1,000	45,000
Total	53		$100,000

There is a total of 53 contracts. That means that for every week in the year an order would have to be forthcoming. There are 43 contracts listed for the $1,000 group. That is a lot of small ones, but one is not going to be bringing in big contracts every week unless he has better than normal connections.

Occasionally we find men who started in electrical contracting with good business connections and all the qualifications of a good contractor. They were able to get started without going through the struggle that is experienced by most beginners. Those are rare cases. That is why the problems of electrical contracting are being discussed here.

As stated before, men with ample capital available have a better chance than those pressed for money. They can carry bigger jobs, put in the necessary stock, and take advantage of cash discounts, and with a good credit rating they gain the confidence of buyers.

Once one has contracts, he must get the men to do the work. We shall next study the man-power problem.

THE LABOR BURDEN

The ratio of the number of mechanics employed to the operating cost serves to emphasize the necessity of having enough work to keep a normal crew steadily employed. Figure 5-1 reveals this in table form.

Before we study the values of Fig. 5-1, two things must be noted: First, the "Annual volume of business" represents the base cost of material and labor. Second, the labor rate used is $1 per hour and material costs are in proportion. Our studies here are based on a labor rate of $3 per hour. To raise the values in Fig. 5-1 to the level of those used in these studies, all items expressed in dollars would have to be multiplied by 3. The percentages and the number of men remain the same. For example, changes in the first line of figures would be as follows:

Col.	Item	Fig. in table	Revised fig.
A	Annual volume—total cost	$10,000	$30,000
B	Annual volume—material cost	$6,000	$18,000
C	Annual volume—labor cost	$4,000	$12,000
D	Operating costs	$4,375	$13,125
E	Operating cost—% total cost (A)	44	44
F	Division operating cost—materials, % of B	15	15
G	Division operating cost	$900	$2,700
H	Division operating cost—labor, % of C	87	87
I	Division operating cost—labor	$3,000	$9,000
J	Number of mechanics	2	2
K	Ratio of mechanics to operating cost, D/J	2,187	6,561

For a further study of Fig. 5-1, the values will be used without adjustment, as they will serve to prove our point. Changing the values would only cause confusion.

BASIS OF EST.
DIV. OF COST – MAT. 60%, LAB. 40%.
LAB. RATE (MECHANICS) $1.00 PER HR.
SALARIES, OFFICE, & GEN. ADMIN. EXP.,
CONSTANT FOR EA. CLASSIFICATION.

LABOR BURDEN
ESTIMATED FOR ELECTRICAL CONSTRUCTION PROJECTS

NOTE:- SUPPLYING & MANAGING LABOR INVOLVES THE MAJOR EXPENSE IN ELECTRICAL CONTRACTING. CHARGES AGAINS MAT. ARE LIMITED BY COMPETITION. THEREFORE, LABOR MUST CARRY ITS RIGHTFUL SHARE OF OPERATING COSTS.

CLASS & VOLUME OF WORK CONTR. IS ORG. TO EXECUTE	ANNUAL VOL. OF BUS. IN DOLLARS: TOTAL COST	DIVISION: MAT.	DIVISION: LAB.	EST. OPER. COST: AMOUNT DOLLARS	% OF TOTAL COST (A)	DIV. OF OPERATING COSTS: MATERIAL % OF B	MATERIAL AMT.	LABOR % OF C	LABOR AMT.	NO. OF MECHS. EMPL. (AV.)	RATIO NO. MECHS. TO OPER. COST (①)	REMARKS	FILL IN YOUR OWN COSTS: OPER. COST AMT.	OPER. COST % OF COST (L & M)	DIV. OF OPER. COSTS: MATERIAL %	MATERIAL AMT.	LABOR %	LABOR AMT.
	A	B	C	D	E	F	G	H	I	J	K							
DWELLINGS-1 TO 12 UNITS	10,000	6,000	4,000	4,375	44	15	900	87	3,475	2	2,187	SEE NOTE 1-BELOW.						
VOLUME UP TO $15,000	15,000	9,000	6,000	4,500	30	15	1,350	525	3,150	3	1,500	⊕ NORMAL CAPACITY						
GROUP 1	20,000	12,000	8,000	4,625	23	12	1,440	40	3,185	4	1,156	TEMPORARY COND.—SEE NOTE 2.						
	15,000	9,000	6,000	7,625	51	15	1,350	104	6,275	3	2,542							
DWELLINGS-1 TO 24 UNITS	20,000	12,000	8,000	7,750	39	15	1,800	74	5,950	4	1,937							
VOLUME UP TO $30,000	25,000	15,000	10,000	7,875	31	12	1,800	61	6,075	5	1,575							
	30,000	18,000	12,000	8,000	27	10	1,800	52	6,200	6	1,333	⊕ NORMAL CAPACITY.						
GROUP 2	35,000	21,000	14,000	8,125	23	10	2,100	43	6,025	7	1,161	TEMPORARY COND.- SEE NOTE 2.						
	20,000	12,000	8,000	10,300	51	15	1,800	106	8,500	4	2,575							
SMALL COMMERCIAL &	30,000	18,000	12,000	10,550	35	12	2,160	70	8,390	6	1,760							
DWELLINGS-12 TO 36 UNITS	40,000	24,000	16,000	10,800	27	10	2,400	52	8,400	8	1,550							
VOLUME UP TO $60,000	50,000	30,000	20,000	11,050	22	10	3,000	40	8,050	10	1,105							
	60,000	36,000	24,000	11,300	19	10	3,600	32	7,700	12	940	⊕ NORMAL CAPACITY.						
GROUP 3	70,000	42,000	28,000	11,550	16.5	10	4,200	26	7,350	14	825	TEMPORARY COND.- SEE NOTE 2.						
	50,000	30,000	20,000	18,250	36	10	3,000	76	15,250	10	1,825							
	60,000	36,000	24,000	18,500	31	10	3,600	62	14,900	12	1,540							
ALL TYPES OF ELECT. CONST.	70,000	42,000	28,000	18,750	27	10	4,200	52	14,550	14	1,340							
JOBS UNDER $15,000	80,000	48,000	32,000	19,000	24	10	4,800	44	14,200	16	1,185							
VOLUME UP TO $120,000	100,000	60,000	40,000	19,500	19.5	8	4,800	37	14,700	20	975							
	120,000	72,000	48,000	20,000	16.5	8	5,760	30	14,240	24	835	⊕ NORMAL CAPACITY.						
GROUP 4	140,000	84,000	56,000	20,500	14.5	8	6,720	25	13,780	28	732	TEMPORARY COND.- SEE NOTE 2						
	120,000	72,000	48,000	43,750	36.5	10	7,200	76	36,550	24	1,820							
	140,000	84,000	56,000	44,250	32	8	6,720	67	37,530	28	1,580							
SELECTED TYPE OF WORK	160,000	96,000	64,000	44,750	28	8	7,680	59	37,070	32	1,400							
LARGE PROJECTS	180,000	108,000	72,000	45,250	25	8	8,640	51	36,610	36	1,255							
VOLUME UP TO $300,000	200,000	120,000	80,000	45,750	23	8	9,600	45	36,150	40	1,146							
	220,000	132,000	88,000	46,250	21	8	10,560	41	35,690	44	1,050							
	250,000	150,000	100,000	47,000	19	6	9,000	38	38,000	50	940							
GROUP 5	300,000	180,000	120,000	48,000	16	6	10,800	31	37,200	60	800	⊕ NORMAL CAPACITY.						
	350,000	210,000	140,000	49,250	14	6	12,600	26	36,650	70	703	TEMPORARY COND.- SEE NOTE 2.						

NOTES:-
① COL. "K" SHOWS THE AMOUNT THAT MUST BE EARNED, FOR EACH MECHANIC WORKING, TO COVER OPERATING COSTS.
⊕ BY WORKING OVER TIME, AN ORGANIZATION CAN TEMPORIARLY HANDLE A VOLUME OF WORK IN EXCESS OF ITS NORMAL CAPACITY

FIG. 5-1. Note: $1.00 has been used as a base labor rate so that the approximate burden for any locality is calculated by multiplying the value in Col. *K* by the prevailing wage rate.

Let us study the figures in the second group, "Dwellings—1 to 24 units, volume up to $30,000" as shown in Fig. 5-1. The fourth line in this group shows the *normal* volume to be $30,000 (Col. *A*). The number of mechanics (Col. *J*) is 6, and the operating cost per mechanic (Col. *K*) is $1,333.

The first line of this same group gives the figures for operating at *half the normal volume*. The volume (Col. *A*) is $15,000, the number of mechanics employed (Col. *J*) is 3, and the operating cost per mechanic (Col. *K*) is $2,542. This represents an increase of over 60% above the normal operating cost of $1,333 per mechanic. No business can long stand to suffer an overload of 60% in operating costs.

COMPARING COSTS

Comparing the cost per mechanic as shown in Fig. 5-1 with those established in our other studies, we find that, in general, they are much greater for small volumes.

In Fig. 3-1, Chap. 3, the minimum operating cost per year was $16,260. With an average crew of five men, this would amount to $3,412 per man. The adjusted (normal) operating cost (see Chap. 4) was $23,835 per year. For a five-man crew, that would be $4,767 per man.

We have two values of studied costs to check:

>   Minimum operating cost per mechanic.... $3,412
>   Adjusted operating cost per mechanic..... $4,767

As stated before, the dollar values in Fig. 5-1 must be multiplied by 3 when they are compared with the values established in these studies, because $1 per hour was used as the base rate of labor whereas $3 is being used here.

In Fig. 5-1, the first line of group 2 gives the figures for a business operating at 50% capacity. The operating cost given in Col. *K* is $2,542. Multiplied by 3 gives $7,626. This is much greater than the studied costs listed above.

The fourth line of group 2 shows the figures for the business operating at capacity. The operating cost per mechanic given in Col. *K* is $1,333. Multiplied by 3 gives $3,999. This is much less than our adjusted cost of $4,767 and approximately 17% greater than the minimum cost of $3,412.

An example of costs that compares closely with the study costs is found in group 2, line 3. The base labor is $10,000. Multiplied by 3 gives $30,000. This is the same as the base labor used in the studies (see Fig. 4-2). The operating cost per mechanic (Col. K) is $1,575, which multiplied by 3 gives $4,725. This is close to coinciding with our adjusted (normal) cost of $4,767.

Not only must a contractor have contracts, but once he has secured them, he must have the proper men to install the work.

CHAPTER 6

Do You Know Where You Are Going to Get the Men?

Do you know where you are going to get five good mechanics? By good mechanics, we mean men familiar with all phases of good electrical construction and who have the push and ability to install high-grade work in normal time. You will need such men if you are going to be able to meet competition and satisfy customers.

An average of five men will be required to take care of a $100,000 volume of work. In the estimate for this amount (see Fig. 4-2), $30,000 was allotted to labor. Figuring 40 hr per week and allowing for holidays, we get approximately 2,000 man-hours per year. At a labor rate of $3 per hour, this would represent a payroll of $6,000 per man-year. The total payroll was $30,000 per year. Dividing 30,000 by 6,000 we get 5, the number of men required if production is normal.

Second to getting work, the man-power problem is the biggest in the electrical contracting business. With some contractors, it is more vital than getting contracts.

Once one is located, his office problem is settled for the life of the lease, he will always find people ready to supply materials and equipment, but the man-power problem is something that is with him at all times. Where they are going to get mechanics is something that does not enter the minds of some men contemplating the contracting business, yet it should be one of the first considerations.

MAN-POWER PROBLEMS VARY

Five mechanics does not seem like a great number to recruit, but when all factors are weighed, the number stands out. Aside from the labor market, one must view his possibilities of procuring work that will prove inviting to good mechanics.

The man-power problem varies with the times, the type of work requiring men, and the community. The effect of the times will be treated later under the heading "Men Available." Here attention will be given to the effect of labor conditions in the individual communities.

The following are some of the conditions of labor that vary in different communities:

1. Open shop—mechanics not organized
2. Closed shop—mechanics belong to a union
3. Open shop and closed shop
4. Apprentice training:
    a. Prescribed course
    b. Hit-and-miss training
5. Seniority board

For those not familiar with all the listed items, some explanation will be provided. It seems hardly necessary to elaborate on the subject of open and closed shop, as one could not grow up in these times without being familiar with it. It might be stated in that connection that most large cities have union electricians. However, a large portion of the electrical work throughout the country is being installed by open-shop men.

The subject of apprentice training is familiar to all, but the methods employed may not be. In some localities, there are no regular rules or regulations for training electricians. A man goes to work for an electrical contractor and, after he has worked with the tools for a while, calls himself an electrician. Most cities of consequence have some regulations regarding training.

Some open-shop as well as closed-shop communities have a graduated scale for men starting in the electrical-construction work. If there is no prescribed course, a man works a certain length of time (usually 4 years) and then is classed as a journeyman electrician.

In general, the better trained men come from localities where there are regular prescribed courses of training which include schoolwork along with the work in the field. In open-shop localities these courses are sponsored by electrical contractors' associations. In the larger cities, representatives of the contractors and the union get together and work out a course of training.

As a rule, the apprentice spends one day a week in school. For one that is inclined to be attentive and study, the school provides an excellent opportunity to learn the fundamentals of physics and electricity. For the most part, the quality of training is assured. The instructors are carefully selected, and the students are required to pass tests.

The fact that a man has completed a 4-year apprentice course does not necessarily guarantee that he is a good mechanic. In the first place, he may not be mechanically inclined and any amount of training would not produce a finished product. In the second place, much depends on the opportunity given him by the contractor he works for and the type of work engaged in.

In spite of the efforts of the union to see that apprentices are given the proper chance, many contractors use them for a year or so as stockroom boys or truck drivers or in other lines of work where there is little to be learned about electrical construction. Most apprentices are started right in helping senior electricians and, by the time they have reached the fourth year, are first-class mechanics. How well they can do on general construction depends on the type of work they have been doing while in training.

The established contractor with large projects can absorb some men with training in limited types of work, whereas the new contractor must have first-class mechanics with all-round training. The former can always release enough first-class mechanics from large projects to take care of the small jobs, but the latter may have nothing but small jobs and will need high-grade men on each.

HIRING UNION MEN

A man expecting to hire union mechanics must meet with heads of the local union and satisfy them that he can meet certain requirements. The approval of the union does not ensure him that there will be men available when needed. It just ensures the contractor that if men are available when needed, they will be permitted to work for him.

A local union may sponsor a seniority board. The purpose of this board is to see that the older mechanics are not pushed aside for newer mechanics or permit men. Permit men are men not regularly belonging to the union but who have been given temporary permits to work as union electricians. Permits are issued when there is an acute shortage of regular union mechanics and withdrawn when work begins to slacken off. The permit men may or may not be good mechanics. As a rule, they are inferior to the regularly employed electricians.

MEN AVAILABLE

When there is an ample supply of good mechanics, work is at low ebb and one does not want to start in business. When there are plenty of contracts to be had, mechanics are scarce. With ample work, the established contractors are careful to hold on to all of their good men. The men, in turn, always want to stay with the established contractors, for they fear that a job with a new contractor may be short lived. Their apprehensions are justified.

A new contractor has no assurance of an even flow of work, and when jobs are finishing up, he has no large projects to put men on and hold them while waiting for more work to come in. However, the established contractor with a large volume of work is often able to absorb excess men temporarily on large projects. The overmanning of the particular job may be costly, but not so much so as letting good men go and then trying to replace them later.

All other things being equal, the new contractor has to take the mechanics that the others do not want. It may take him 2 or 3 years before he has a well-rounded crew.

The reader may remember that in the estimate of the cost for getting started, no allowance was made for losses due to the use of inferior mechanics. This item may be appreciable.

COST OF INFERIOR MECHANICS

Electrical Estimating (McGraw-Hill Book Company, Inc.) defines productivity (man power) as *the ratio of the actual man-hours consumed on a job to the standard man-hours for a similar project.*

For several years the labor records of 1938 have been used as standards of man-power productivity. Two reasons brought this about: (1) Extensive studies of productivity were made in 1938. (2) It is much simpler to establish reliable data for one year than it is for several.

The 1938 records provided a good gauge, but the reader must be reminded that in so far as the mechanics themselves were concerned, their productivity rating was above average. Only the better men were employed, there were few apprentices and no permit men. Compared with the average of the previous 10 years, the productivity of 1938 was more than likely about 120%.

Although the productivity of the average mechanic, today, is less than that of 1938, the amount of work turned out per mechanic should not be much less. Increased use of tools and improved types of tools and construction equipment have greatly increased the amount of work that a mechanic can turn out.

To simplify the examples used here, we shall assume that the crew of the established contractor averages 100% productivity. The rating of the new contractor's crew will be expressed as a percentage of the established contractor's. To illustrate the effect of low productivity, values will be assigned to the new contractor's crew and the results figured.

For the first example, 80% will be used as the productivity rating of the new contractor's crew. An 80% crew will take 25% longer to complete a job than will a 100% crew. To illustrate the penalty of this low rating, estimates will be made for a $10,000 base-cost project with an M/L ration of 60/40.

Example 1a. Normal labor (100%)

	Material		Labor	
Base costs (purchase price materials and payroll)	$ 6,000	60%	$4,000	(40%)
Direct job costs (not including insurance) .... (2%)	120	6%	240	
	$ 6,120		$4,240	
Overhead 10%	612	35%	1,484	
Total cost—material	$ 6,732		$5,724	
Total cost—labor	5,724			
Total cost—the job	$12,456			

Example 1b. Labor 20% below normal (80%)

	Material		Labor	
Base costs, material and labor—normal	$ 6,000	60%	$4,000	(40%)
Excess labor			25%	1,000
Experienced base costs	$ 6,000		$5,000	
Direct job costs (not including insurance) .... (2%)	120	6%	300	
	$ 6,120		$5,300	
Overhead 10%	612	35%	1,855	
Total cost—material	$ 6,732		$7,155	
Total cost—labor	7,155			
Total cost—the job	$13,887			

In Examples 1a and 1b, the total labor costs are $5,724 and $7,155, respectively. The excess labor burden of 25% (due to 80% productivity) expressed in dollars is $1,431 (7,155 − 5,724). Since there was no change in material costs, the difference in the total job costs is also $1,431.

The increase of $1,431 is equal to 11.5% of the normal job cost (see Example 1a) of $12,456. On the estimated normal volume of $100,000, this 11.5% would amount to $11,500 per year. Expressing it in terms of productivity, we can say that the penalty for 80% labor productivity represents an annual burden of $11,500.

Before we go on with the discussion, another example will be studied, using 90% as the factor of labor productivity.

A crew with a 90% rating would take 11% longer to complete a job than a crew with a 100% rating. Employing the $10,000 base cost and the 60/40 material-labor ratio used in the previous examples, the figures would be:

Example 2a. Normal labor

The same as Example 1a.

$$\begin{array}{ll} \text{Total material cost} & \$\ 6,732 \\ \text{Total labor cost} & \underline{5,724} \\ \text{Total cost—the job} & \$12,456 \end{array}$$

Example 2b. Labor 10% below normal

		Material		Labor
Base costs, material and labor—normal		$6,000	60%	$4,000 (40%)
Excess labor			11%	440
		$ 6,000		$4,440
Direct job costs (not including insurance) (2%)		120	6%	266
		$ 6,120		$4,706
Overhead (10%)		612	35%	$1,647
Total cost—material		$ 6,732		$6,353
Total cost—labor		6,353		
Total cost—the job		$13,085		

$$\begin{array}{ll} \text{Example 2b. Total cost—the job} & \$13,085 \\ \text{Example 2a. Total cost—the job} & \underline{12,456} \\ \text{Excess burden (2b)} & \$\quad 629 \end{array}$$

The excess labor cost in Example 2b is $629. This is better than 5% of the normal job cost of $12,456. Again looking at the burden in terms of the $100,000 annual volume, the 5% represents $5,000 per year.

To say that a crew has a 90% rating does not sound serious, but when viewed in terms of the business, it is bad. Although the $629 burden is only 5% of the total job cost, it is 50% of the profit (10%)

that one would normally expect to enjoy on a contract of that size.

The reader may find himself asking, "But why does the owner hire so many men? Why doesn't he work with the tools himself?"

In that case the reader must remind himself that we are not studying a one-man business where the owner is more or less a free-lance workman. We are studying a one-man business that is expected to grow and keep on growing. It should graduate into the $500,000 to $1,000,000 class in the course of 4 or 5 years. However, no matter how small the business, if one has to depend on hired help at all, he will have man-power problems.

To keep going and growing, the contractor must select his work judiciously. To do this, he must be able to appraise projects from the standpoint of desirability.

CHAPTER 7

Are You Familiar with the Functions of Contracting?

Can you estimate?
Can you engineer work?
Do you know how to select the better jobs?
Can you direct work in the field?
Do you know how to figure operating costs?
Can you sell?

The above listing shows some of the major functions of good electrical contracting. To this list can be added many more. Too much weight and consideration cannot be given to any of them.

CAN YOU ESTIMATE?

A good knowledge of estimating is a must requirement in the electrical contracting business. A little knowledge of estimating may be bad, and tables of labor units do not enable one to be an estimator. Do you know that each job must have its individual estimating units? You can no more create tables of labor units that will be applicable to all jobs than you can make a suit that will fit all comers.

Do you know that markups suitable to various types and sizes of jobs are as individual as labor units? When the subject is treated in a later chapter, it will be seen how vital the selection of the proper markups is to good business.

One can learn a great deal by studying the subject of electrical estimating, but to be a finished estimator he must have experience in the actual work. He must also be trained in the study of labor costs and the preparation of labor unit tables. The subject of learning to estimate is too extensive to be covered here. It is treated at length in the book *Electrical Estimating* (McGraw-Hill Book Company, Inc.).

Some men think that they will hire trained estimators and thereby get along without any special knowledge of how to estimate work. There are many reasons why this is not a solution. In the first place, as asked before, where is the money coming from to pay this man while waiting for contracts? Suppose one had the money to pay an estimator, how sure would he be that he was going to get the right man?

One may be able to hire a good estimator, but his estimating problem would still not be settled. As long as one has to depend on someone else, he will be groping in the dark. He must know enough about estimating to be familiar with the work his estimator is preparing.

A contractor may hire ever so good an estimator, but unless he knows estimating himself, as soon as his competitors begin to get all the work, he will lose confidence in his estimator. He must be able to check the estimates and convince himself that they are right.

Do Not Let Your Competitor Do Your Estimating

Contractors who are not sure of themselves are prone to let their competitors set their prices. If they consistently lose jobs, they begin to lower their labor units and markups. Before they know it, they have a losing job on their hands. As long as there has been electrical contracting, there have been men going through that cycle. They consistently lose contracts, cut prices, get a losing job, temporarily stiffen up on prices, and then start the cycle all over again. All the while profits from regular customers keep them in business.

If you ever feel that you must get a job and that the only way to do it is by reducing prices, just be sure that you know what you are doing. Do not take off a little here and a little there and then take off some more at the end of the estimate. Figure the work as you normally would in close competition, and use your established markups. After the estimate is completed, start the paring. When you enter your bid, you will know how much of a sacrifice you are making.

When consistently losing contracts, before starting to make changes in prices, study your estimates and review prevailing conditions. You may not have allowed for improving labor market or the possibility of being able to buy more advantageously.

Electrical estimating involves much more than just taking off quantities and pricing material and labor. One must always be alert to changing times and be able to recognize the hazards or special advantages of each individual project. Many contractors have lost jobs because their competitors studied the work more carefully and were able to think up more ingenious methods for installing the work.

DO YOU HAVE A KNOWLEDGE OF ENGINEERING?

One need not be an electrical engineering graduate to be an electrical contractor. However, such training is an asset. On the other hand, being a graduate electrical engineer does not mean that one is versed in electrical-construction engineering.

Very few colleges offer courses designed to aid students who wish to enter the electrical contracting field. Yet today there are thousands of college graduates working for electrical contractors or architects and engineers designing electrical installations.

Many sons of electrical contractors and others enter college with the idea that they will become affiliated with electrical contracting after graduation. Instead of finding something adapted to their purposes, they are obliged to take courses designed to train men for large manufacturers of electrical equipment, utilities, or other industries little akin to electrical contracting.

Many contractors started in the business with a limited amount of schooling but, by taking advantage of correspondence courses or night schools or by other means, were able to get a fair knowledge of electrical engineering. While they studied, they were in contact with construction work.

In the early days of electrical contracting, few men engaged in it were well educated. As the industry grew, they grew with it. Their engineering abilities were limited, but they had a good hold on the business and were able to augment their knowledge with that of good engineers in their service. They had the feel of the business and could rely on others with impunity.

From the standpoint of engineering, the contractor is confronted with two types of projects, namely:

1. The project that has well engineered and detailed plans and good specifications
2. The project that must be engineered and laid out

No matter how well a job is engineered, laid out, and specified, the contractor must have some knowledge of engineering in order to figure it and plan the installation. He must understand the hookups of motors, controls, transformers, and other types of equipment. He must also be able to design and select construction equipment such as pull boxes, platforms, hangers, etc.

For jobs that have to be laid out, the knowledge of engineering required depends on the size and type of installation. On a simple job the engineering may be limited to the selection of standard materials and the figuring of the size of lamps and feeders. However, for large projects the engineering problems are usually numerous and involved.

On very large projects one must determine the sizes and designs of services, feeders, transformers, load centers, etc. For an example, let us take the lighting system. The intensity of light that will be required must be determined, and the type and size of fixtures settled. The feeder problems involve the voltage and type of current. Should it be high voltage with transformers and tap-offs at the panels or low voltage with panels connected directly to the feeders? If transformers are to be used, the type, size, spacing, and method of installing must be determined.

The installation of lighting transformers requires many detail studies. The building construction must be studied to determine a suitable location, and the method of supporting designed accordingly. Grounding details must be provided giving size of wire, type of ground terminal, supports, and guards. The sizes of wire for the primary and secondary connections must be figured, and the proper conduit and fittings selected.

The selection of the type of current to be used for lighting feeders is dependent on the service available. The final selection may be two-wire single-phase, three-wire single-phase, three-wire three-phase (grounded transformer tap-offs), or three-phase four-wire.

The larger the project, the greater the number of essential constuction details. Locations of feeders and equipment must be established and almost any part of the installation may require detailed drawings.

GOOD PLANS A MUST REQUIREMENT

Well-laid-out plans are a must requirement in good electrical-construction practice. Any smart contractor will want to prepare good plans before the job is started, and any well-informed owner will insist on having them. The better contractors know that it is economy to spend money to have installation plans made. There will be more on this subject under the heading "Management."

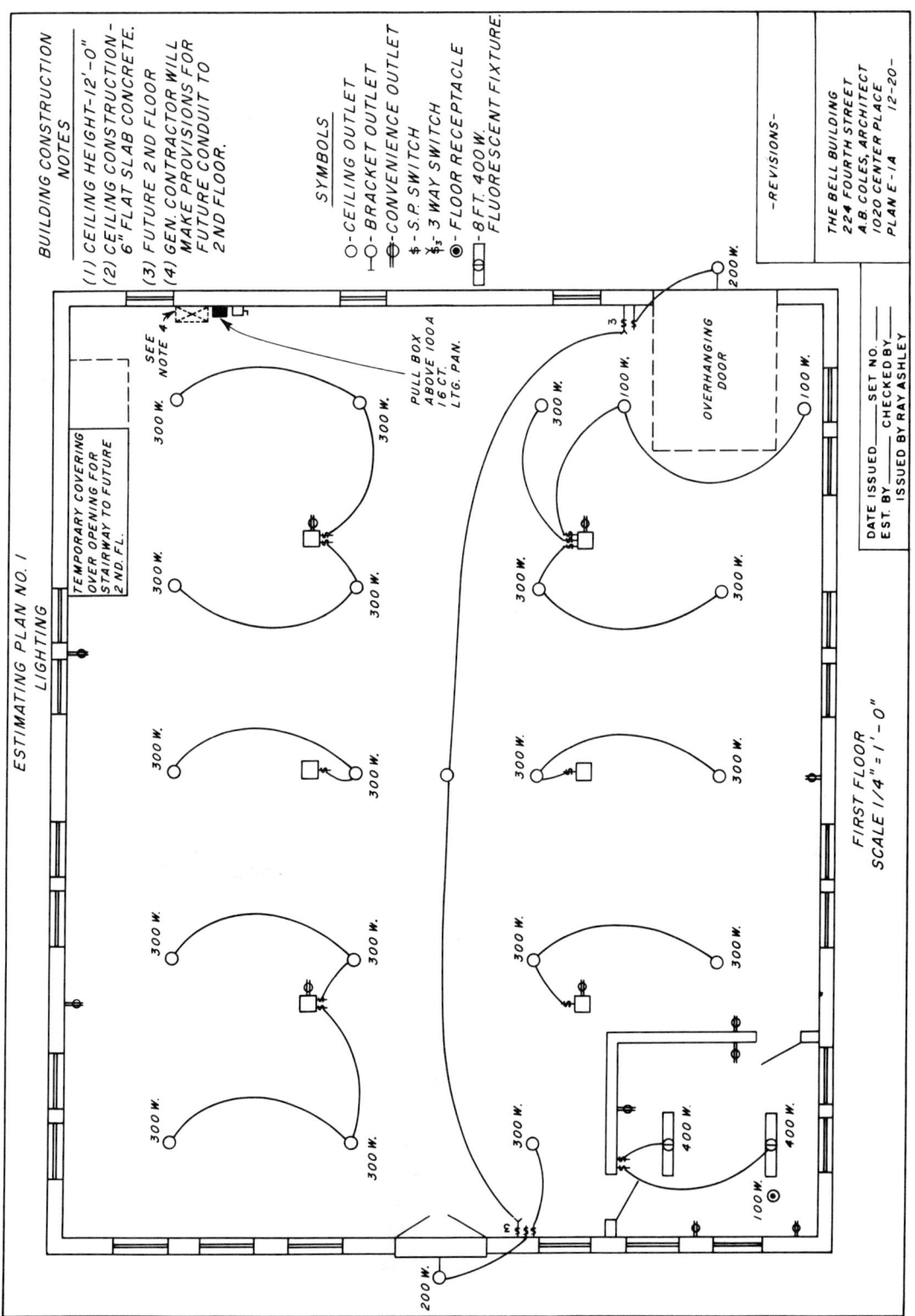

Fig. 7-1. Lighting plan with outlets and switches only shown. Contractor must complete layout.

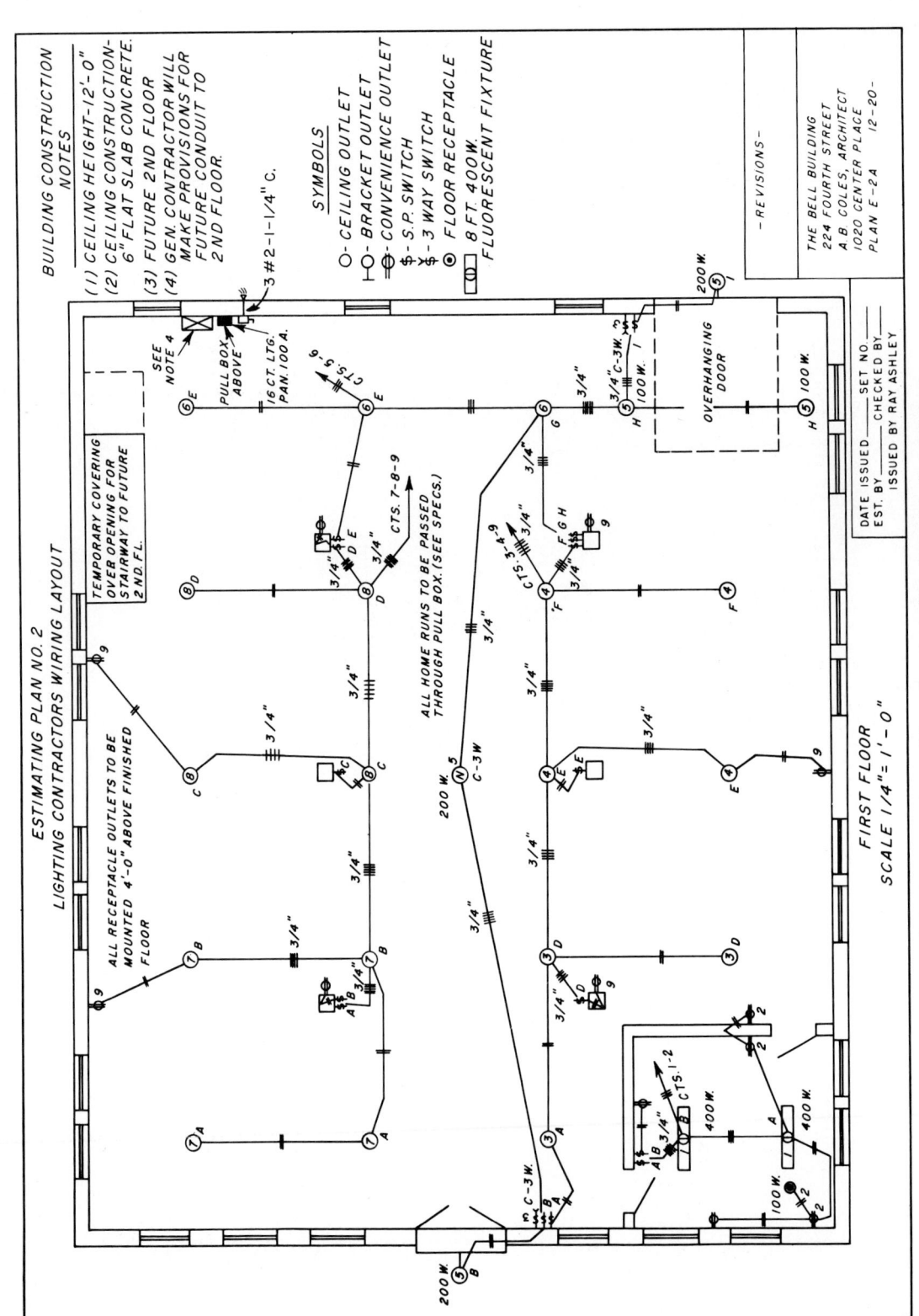

Fig. 7.2. Contractor's working drawing. See Fig. 7-1 for original.

Figure 7-1 shows a lighting plan with outlets located and switching indicated. Figure 7-2 is a plan for the same lighting, showing the type of layout that must be made by the firm having the contract for the installation.

Figure 7-3 is a preliminary power plan giving the size and layout of motors. Figures 7-4 and 7-5 are the contractor's plans for the installation of same. The final plans include the specification requirements.

It is unfortunate that not all plans laid out by electrical contractors are well engineered. Perhaps we should say that not all electrical layouts are well engineered. Many installations laid out and engineered by plant engineers and some men who call themselves professional engineers do not reflect high standards of enginering. For the buyer, employing a good reliable contractor with engineering ability is like buying high-grade insurance.

All other things being equal, the contractor with a good reputation as an electrical and construction engineer has the advantage over his less competent competitors. He is called in for counsel by architects, engineers, and owners and is often in position to close contracts without having the plans thrown out for free-for-all bidding.

As stated before, there will be more about well-laid-out plans later. At that time the economy will be studied.

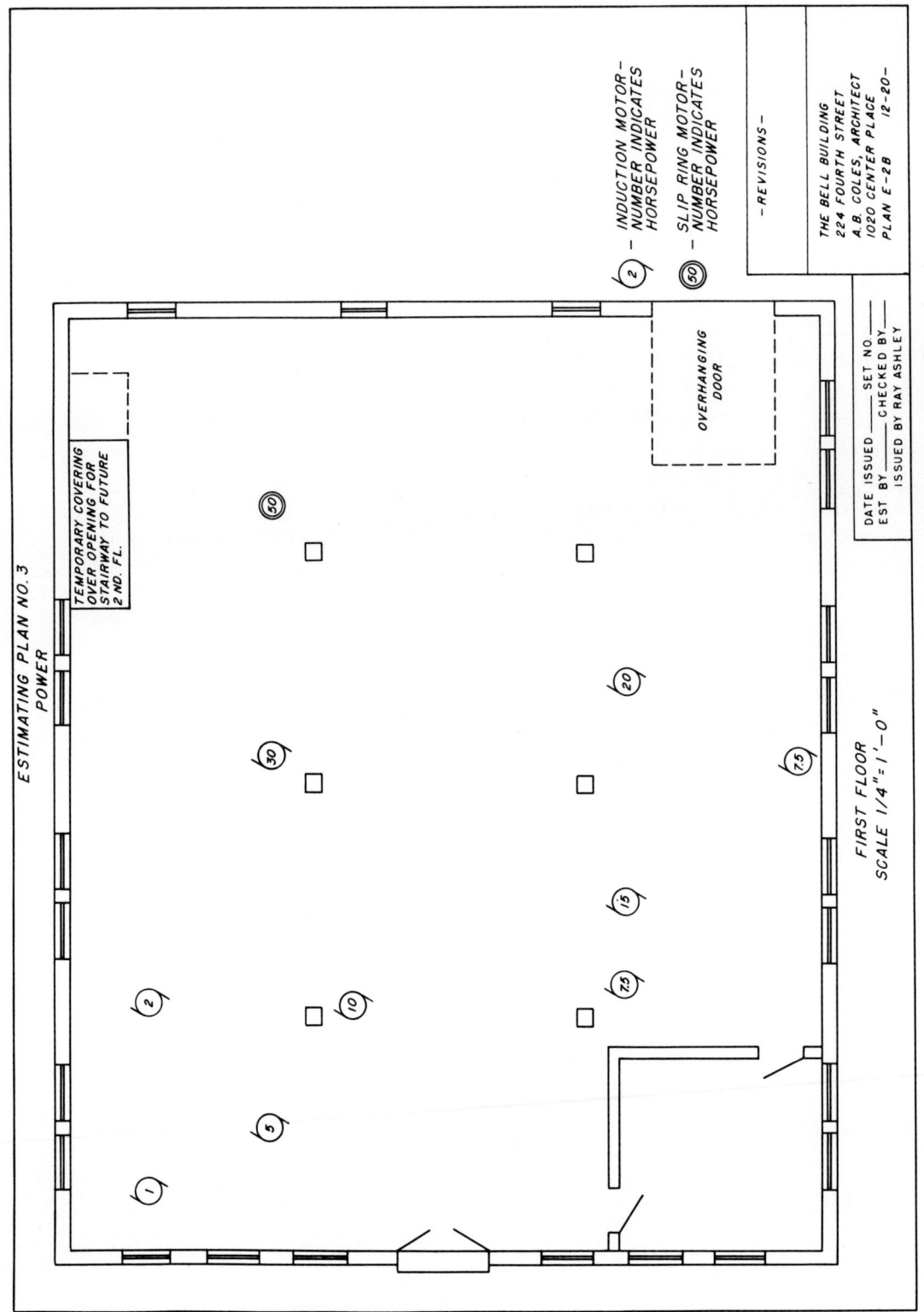

Fig. 7-3. Plan showing size and location of motors. Contractor must complete drawings.

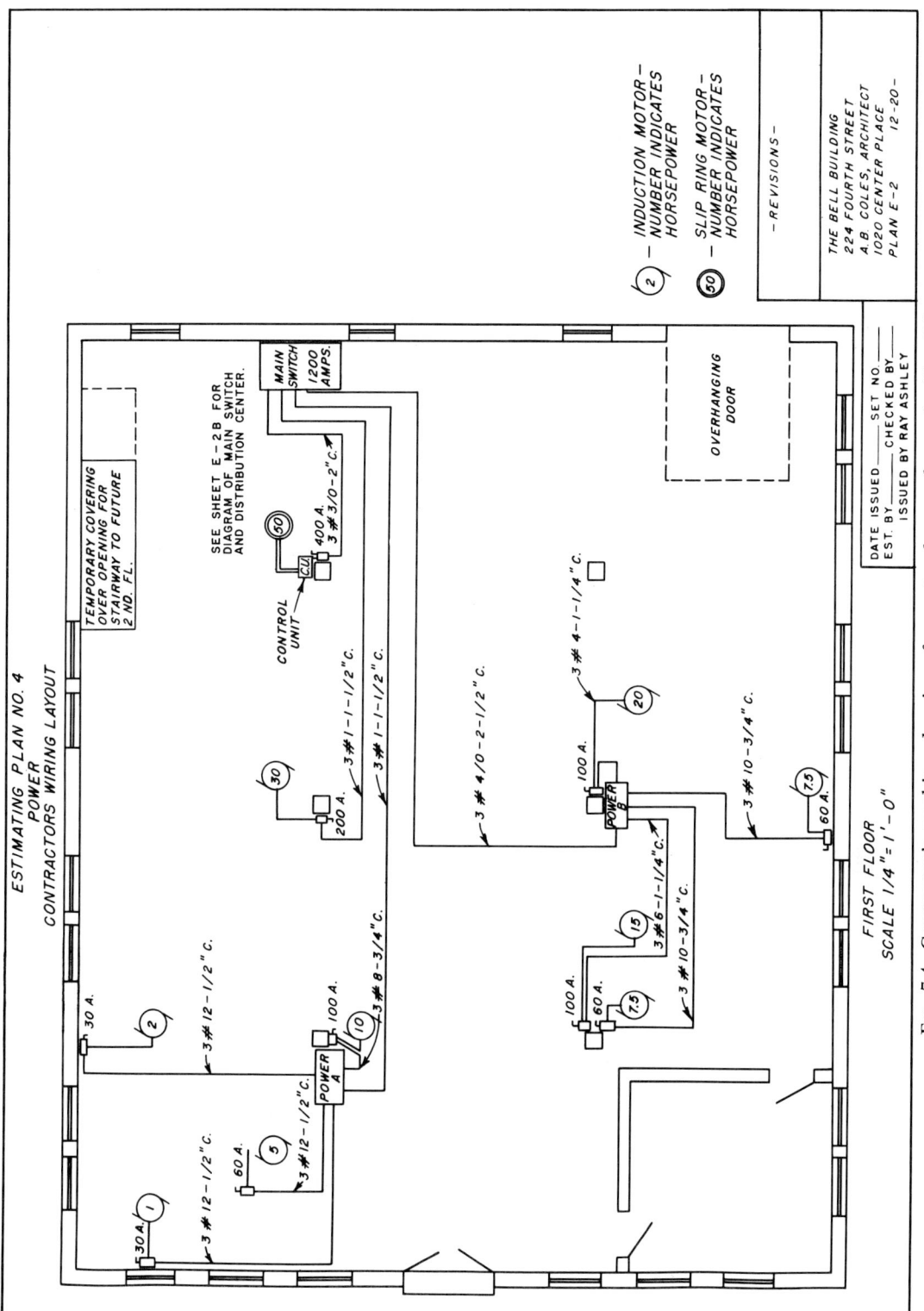

Fig. 7-4. Contractor's working drawings for motor layout shown in Fig. 7-3.

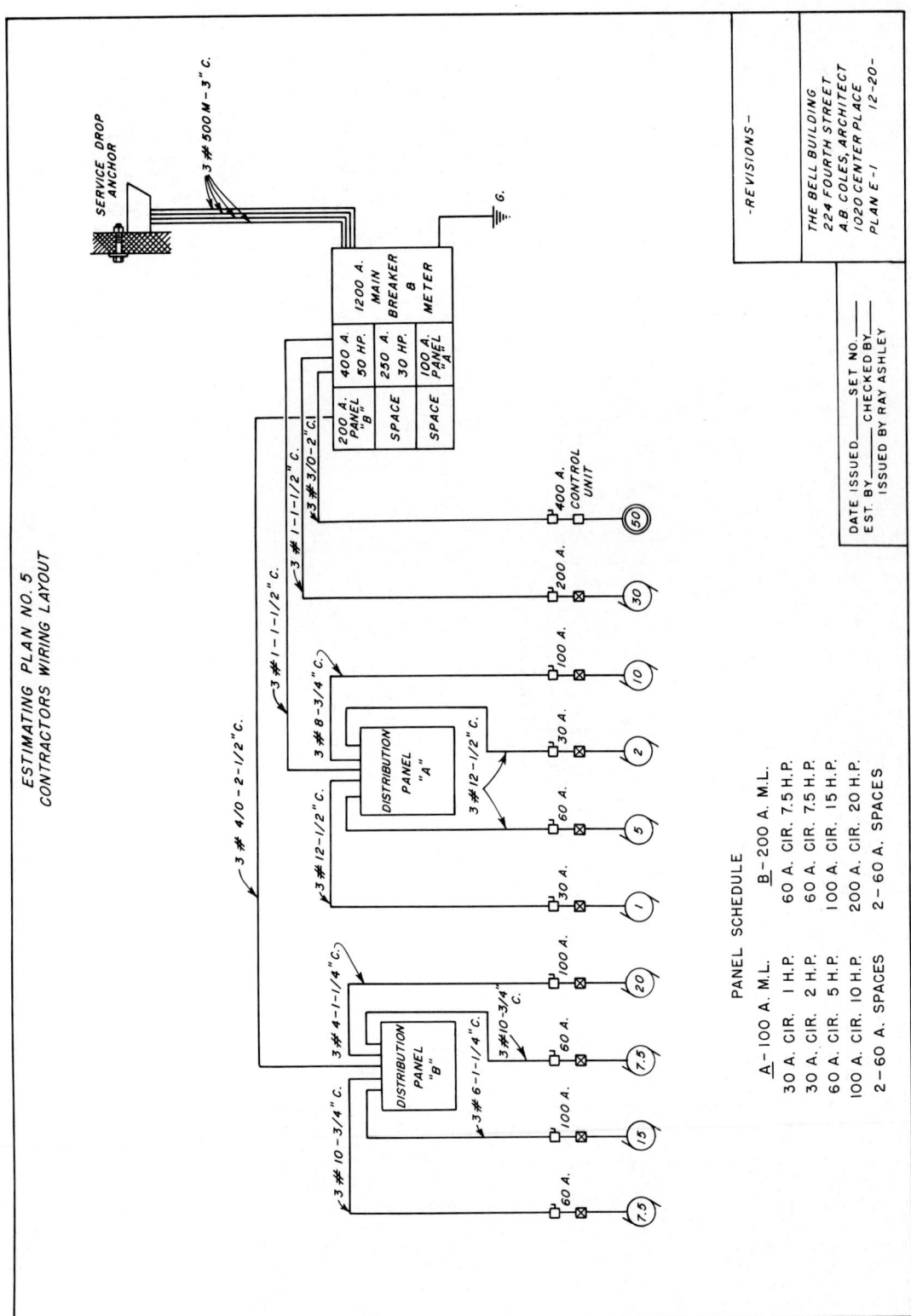

Fig. 7-5. Contractor's power "riser."

FIG. 7-6. Contractor's take-off for power.

CHAPTER 8

Can You Select the Better Contracts?

Do you know that the volume of work handled by a contractor can vary as much as 200%, depending on the type of contracts secured? Do you know that a job that runs twice as long as it should costs 10% more than it would for normal duration? And do you know that installation-only (material by owner, installation only by contractor) projects are not generally so desirable as complete installation contracts? These are only a few of the many questions that can be asked in order to point out that there can be a vast difference in the desirability of contracts of the same dollar volume.

No attempt will be made here to classify many types of work. To do so would be a big undertaking, and the discussion would prove irksome to the reader. However, to illustrate how much the desirability of contracts can vary, two distinct types of installations will be studied. For comparison, an alteration project and an industrial feeder installation will be selected. Each will be elaborated on in turn.

ALTERATION WORK

As a rule, small and medium-size alteration jobs are not considered desirable. For the amount of money involved, the cost of estimating is high, supervision is costly, duration is prolonged, and the turnover of investment is slow. On alteration work mechanics and equipment are tied up for a long time.

A phase of alteration work that justifies particular attention is that of the potential hazards. Most of these are well known to seasoned estimators and experienced contractors, but there are three in particular that are often overlooked by beginners, namely:

1. Progress of other trades
2. Occupied spaces
3. Concealed work

Progress of Trades. The work may have to be done piecemeal because one trade works a while and then waits for another to catch up. Other trades may have considerable work, whereas the electrical contractor can put in only a small amount of work from time to time so that his wiring spaces will not get covered.

In the case of suspended ceilings, the electrician has to work along with the lather. On large alteration jobs, the effect of working with the lather is not different from that of a new installation. For small alteration work, the progress of the lather may be very irregular.

Occupied Spaces. The hazards of occupied spaces vary according to the type of installation. There are offices, factories, warehouses, hospitals, and many other types of installations, each of which has its own problems. In a warehouse, it is a case of moving or working around stock. In an office, furniture must be moved and protected, noises limited, and working spaces kept clear and clean at all times. In a factory, men have to work over and around moving machinery.

An entire chapter could well be devoted to the subject of work in occupied spaces. Here the discussion must be limited. In general one may expect to find labor increases above normal, for alteration work in occupied spaces, to be approximately as follows:

$$\begin{array}{ll} \text{Offices} & 30\text{--}50\% \\ \text{Factories, over machinery} & 30\text{--}60\% \\ \text{Factories, over benches} & 25\text{--}35\% \end{array}$$

Concealed Hazards. Contractors, estimators, and mechanics soon learn the seriousness of hazards on jobs where existing walls have to be removed. Specifications for such work usually hold the contractor responsible for all concealed work. He must move conduit and wire and keep the system operating at all times.

There is one type of restoration work that the novice must approach with caution. That is the fire job. One is inclined to assume that the damage is limited to the burned-out section and that the damaged wire can be easily removed because it is loose at that point.

The wire will be loose in the conduit at the burned-out section because the insulation is off. There will be other sections where the heat was just great enough to vulcanize the rubber and the wire will be frozen in. This may happen some distance from the actual fire. The pipes get red hot, and when the water is poured on, the heat shifts. We have all had the experience of inserting the heated end of a rod in water and having the heat rush to the end we are holding.

INDUSTRIAL PROJECTS

In a previous paragraph, it was stated that an alteration project would be compared with an industrial installation consisting of power feeders. We have studied the former; here our attention will be turned to the latter.

All types of industrial work where the contractor supplies the equipment and material are sought after by contractors. Installation of large power feeders can be especially desirable. On such installations, the contractor has complete control of his work and is not dependent on the progress of other trades. The turnover is rapid, there are no concealed hazards, mechanics can be used to an advantage, and the M/L ratio is especially good.

Quick Turnover. The quick turnover on a job means much more to the electrical contractor than just the turnover of the money invested. It means that good foremen and tools are not tied up as long as they would be on many other types of jobs of the same volume. A contractor's capacity is limited by the number of good foremen and superintendents he has available. In fact, the whole organization functions more effectively on a power feeder job than it does on an alteration project. One may be able to do more than twice the volume if contracts can be of the select instead of the hit-and-miss type.

Mechanics Required. The man-power problem is not so serious for industrial work as it is for other types of installations. In the first place, a foreman can direct work more easily, and in the second place, more second-rate mechanics can be absorbed.

The interest mechanics take in their work is greater on industrial projects than it is on many other types. It is only natural for a mechanic to take more interest in work that he can follow to completion without interruption. He likes to see what has been accomplished.

On a multistory concrete building, one installs conduit and boxes which are soon covered with concrete. He stubs up conduits and goes on leaving them to be extended at a later date. When it comes time for trimming the job, he may be working in a section entirely new to him. Under such conditions, one does not have the same incentive that he has when able to do a piece of work completely.

Possibility of Future Work. The possibilities of one job in an industrial plant leading to another are good. Most industrials are constantly making additions or changing departments around. The contractor who does his work well is paving the way for future calls from the owner.

Many contractors have found that during dull periods there was more work coming from their regular industrial customers than any other sources. At such times a small amount of work does more than provide income; it is a great tonic for a sinking morale.

OTHER TYPES OF CONTRACTS

Although industrial work has been used as an example, there are many other types that have been found interesting and remunerative. Much money has been made on office and commercial buildings. Schools, hospitals, theaters, and many other types of buildings have provided remunerative work for contractors who understood them.

The competition is greater for industrial work than it is for run-of-the-mine jobs. It is easier to figure and offers less hazards. Contractors complain about the competition in the fields of hospitals, office buildings, and many other types because they do not like to figure so closely on work that, at best, is found to present an element of uncertainty. In spite of the complaints, the owners always find plenty of bidders.

WHAT CONSTITUTES A DESIRABLE JOB

Two types of contracts were discussed at length to illustrate the factors that may make one job more desirable than another. The fact that the more desirable factors were included with the industrial job and the less desirable in the alteration class does not mean that, in general, either is desirable or undesirable.

A desirable project is one the performance of which will yield a profit in both money and satisfaction. Work that is desirable for one contractor may not be good for another. Again, the cycle of an individual business changes and work that would be beneficial for a contractor at one time might not fit in his program at another. Competition can quickly change the class of work that the individual contractor finds worth figuring.

We enter business with the idea of making money. Naturally, competition is a limiting factor. We must select work to figure that is in a line where we can meet competition.

FACTORS TO BE STUDIED

The first consideration of any contractor, when contemplating a job, must be his ability to carry on the work properly. That being settled, the following are factors to be taken into account:

1. Cost of estimating
2. Cost of engineering
3. Hazards involved
4. Material-labor ratio
5. Duration
6. Effect on existing work
7. Possibility of leading to future business
8. The general contractor
9. The architect
10. The competition

Cost of Estimating and Engineering

Along with estimating always goes some engineering. Here, when we speak of estimating, it will be understood that the engineering is included.

The total cost of estimating must be prorated to the contracts secured. Generally the cost of estimating is greater for work obtained from general contractors than it is for work obtained from architects and owners, the principal reason being that the general contractors take more bids and the electrical contractor's chances of getting a job are less per estimate.

Let us say that it costs $100 to estimate a job. If three bids were taken by the architect or owner, the contractor's chances of getting a job would be one out of three and the total cost of estimating to get that job would be $300. By the same method of figuring, if general contractors took bids from 10 electrical contractors, the cost of estimating per contract obtained would be $1,000.

Some jobs in themselves are so difficult to figure that with keen competition one is not justified in gambling the estimating time.

Hazards Involved

Most contracts have some potential hazards. It was pointed out before that alteration work can be especially hazardous. There are always general hazards in construction work such as uncertain labor market, strikes, job shutdowns, and rising prices. All through these chapters, problems will be discussed which represent potential hazards. Any-

CONSTRUCTION PERIODS FOR ECONOMIC OPERATION OF ELECTRICAL INSTALLATIONS — INDUSTRIAL
SEE NOTES BELOW

LABOR		CONSTRN. PERIODS-SEE SKETCH BEL.									TOTAL PERIOD		AV. NO OF MEN EMPL
MAN HOURS	MAN DAYS	1 & 6			2 & 5			4 & 3					
		DAYS	MEN AV.	MAN DAYS	DAYS	MEN AV.	MAN DAYS	DAYS	MEN AV.	MAN DAYS	DAYS	WKS.	
1	2	3	4	5	6	7	8	9	10	11	12	13	14
1,250	156	10	2	20	12	4	48	15	6	90	37	7.4	4.2
3,150	394	12	2	24	35	6	210	16	10	160	63	12.5	6.3
6,400	800	12	2	24	52	8	416	30	12	360	94	19.	8.5
13,600	1700	15	4	60	75	12	900	37	20	740	127	25.4	13.4
28,000	3,500	20	4	80	125	16	2,000	60	24	1,440	205	41	17.1
44,000	5,500	30	5	150	145	22	3,190	60	36	2160	235	47	23.4
59,500	7,440	30	5	150	160	26	4,160	70	45	3,150	265	53	28.5
75,000	9,375	35	5	175	175	30	5,250	75	53	3,975	285	57	33.
90,000	11,250	40	5	200	185	34	6,290	80	60	4,800	305	61	37.
112,500	14,060	42	7	294	225.	37	8,325	85	64	5,440	350	70	40.
150,000	18,750	45	11	495	270.	45	12,150	85	72	6,120	400	80	47.
190,000	23,750	50	14	700	295	54	15,930	90	80	7,200	435	87	54.5
230,000	28,750	60	15	900	300	66	19,800	90	90	8,100	450	90	64.
267,000	33,375	65	16	1,040	304	72	21,888	95	110	10,450	465	93	72.
305,000	38,125	70	18	1,260	310	79	24,490	100	125	12,500	480	96	80

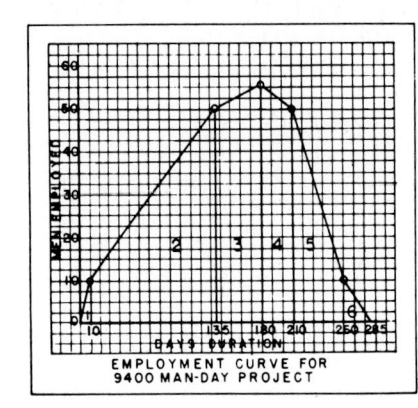

ESTIMATED OPTIMUM DURATION FOR A JOB REQUIRING 9400 MAN-DAYS

PERIOD	DAYS	NO. OF MEN-AV.	MAN DAYS
1	10	5	50
2	125	30	3750
3	45	53.	2385.
4	30	53	1590
5	50	30	1500.
6	25	5	125.
TOTALS	285		9400

EMPLOYMENT CURVE FOR 9400 MAN-DAY PROJECT

1.—PERIODS BASED ON PROGRESS FREE OF INTERFERENCE FROM OTHER OPERATIONS. 2.—CONTRACTOR MUST BE MANNED, EQUIPPED AND ORGANIZED TO EXPEDITIOUSLY CARRY ON THE WORK.

FIG. 8-1. The table readily reveals the approximate man-power demands and optimum duration for contracts with various estimated man-days.

thing that may cause excess expense or cause the work to get out of hand is a hazard.

Material-Labor Ratio

Studies show that on a dollar basis, it costs less to supply material than it does to supply the labor and labor services to install it. For this reason work with a high ratio of material is generally considered more desirable than work with less material and more labor.

Installation-only projects (material by owner, labor and labor services by contractor) are usually considered less desirable than complete installations, the two principal reasons being that the hazards are greater and the contractor's volume is reduced. A contractor is equipped and organized to take care of supplying the material along with labor. If he supplies the labor and labor services only, with the same organization, he will be able to do less than one-half his normal volume.

Installation-only projects are a study by themselves and will be treated later.

Duration

Under the subject of "Industrial Projects," short duration was listed as one of the favorable factors. All installations have an optimum duration period. Figure 8-1 lists the desirable duration periods for contracts within a given range. Smaller contracts also have an optimum duration.

In selecting contracts, one must consider the length of time key men and construction equipment will be tied up. Slow-moving work limits the possible volume. For estimating costs of extended duration see *Electrical Estimating* (McGraw-Hill Book Company, Inc.).

Effect on Existing Work

A contractor must estimate and try to get contracts that will fit in with the work he has in progress. As one project nears completion, there must be another opening up to absorb the men. An even flow of work is a must item if the contractor's men and equipment are to be used to the best advantage.

Future Business

Work that seems to be of the nuisance type often turns out to be a real asset. It may lead the way to some valuable work in the future. Repair work in itself is not generally desirable, but it may pave the way for getting future contracts that are remunerative.

REPAIR SERVICES
(Electrical)

STUDY OF OPERATING COSTS FOR BUSINESS AS FOLLOWS

4 Mechanics (Av.) Employed 40 Hrs./Wk. $3.80/Hr. Av. Wage

Div. of Base Cost – 30% Mat. (pur. pr.), 70% Lab. (payroll)

ANNUAL VOLUME – BASE COST
- Payroll $31,620. – 70%
- Material 13,550. – 30%
- TOTAL $45,170.

OPERATING COSTS

ITEM	REMARKS	COST PER YEAR	DIVISION MATERIAL	DIVISION LABOR
ADMINISTRATIVE				
SUPERVISION				
ENGINEERING	By Prop. $200./Wk.	$10,400.	2,080.	8,320.
ESTIMATING				
PURCHASING				
SELLING				
STORE ROOM ATTENDANT				
BOOKKEEPING – GENERAL				
BOOKKEEPING – SP. TAXES	$75./Wk	3,900.	1,170.	2,730.
Steno. & Office Attendant				
RENT – Office	$100./Mo.	1,200.	360.	840.
RENT – Store Room				
LIGHT	$15./Mo.	180.	54.	126.
HEAT		300.	60.	240.
TELEPHONE	$20./Mo.	240.	48.	192.
OFFICE FUR. & EQUIP.	0.25% of $45,170.	113.	23.	90.
STR. RM. Bins, Rks. Etc.				
STATIONARY & SUPPLIES	0.50% of $45,170.	226.	46.	180.
POSTAGE	0.40% of $45,170.	181.	36.	145.
TAXES & LEGAL EXP.	0.25% of $45,170.	113.	23.	90.
ADVERTISING & DONATIONS	0.20% of $45,170.	90.	18.	72.
COLLECTING BAD DEBTS	0.10% of $45,170.	45.	11.	34.
INT. on PAYROLL	0.50% of $31,620.	158.	–	158.
TRAVEL EXPENSES	Light Truck	1,200	420.	780.
CARTAGE				
TOOLS – Consumed & Depr.	2% of $31,620.	632.		632.
MISCELLANEOUS		200.	40.	160.
TOTALS		19,179.	4,389.	14,789.

TOTAL OPERATING COST – 42.4% of Combined Labor & Material Cost.
Material SERV. COST – 32.4% of Material (Pur. Price)
Labor SERV. COST – 46.7% of Labor (Payroll)

NOTE: Insurance included in payroll.

Fig. 8-2. Repair-service operating costs vary greatly from those for regular contract work. This study is for a wage of $3.80. A locality with any other rate would have much the same percentages of operating costs.

Figure 8-2 provides a study of the cost of supplying repair services. The material-labor ratio is low, and the operating costs high. It is hard to convince buyers that such markups are normal. Often contractors have to absorb some of the repair-service costs in order to avoid antagonizing regular customers.

The General Contractor

Some general contractors expedite their work better than others, and all the subtrades benefit. The cost of roughing-in electrical work may vary 10% or more, depending on the firm handling the general contract.

In selecting work to figure, an electrical contractor is foolish to go to a general contractor unless he knows he will be given a fair chance. We hear a great deal of complaint from the subtrades about the abuses from general contractors. Most of us know that much of the abuse is invited by the subtrades.

The Architect

There are questions a contractor must settle regarding an architect before appraising the value of the contracts he may have to let. The following three are important:

1. How well are his jobs engineered?
2. Which general contractors is he likely to employ?
3. How well does he expedite his work?

Before soliciting an architect's work the contractor must decide whether or not it is desirable and whether or not the architect considers him favorably.

A CASE OF GOOD JUDGMENT

We have outlined certain factors that earmark some jobs as being better than others, but there are no fixed rules for proving the value of work. Any job is good if it will yield a return commensurate with time and services invested.

It takes study, experience, and good judgment to enable one to select the better contracts.

CHAPTER 9

Can You Make the Most of Contracts?

Once you have secured a contract, can you do the job well and realize the greatest possible gains for your efforts? Do you know that one contract properly handled may earn more than two similar contracts run in a slipshod manner? The following are some of the questions to be answered relative to the management of contracts:

1. Can you select men properly?
2. Can you direct men effectivly?
3. Can you recognize the job needs for plans and shop drawings?
4. Are you capable of engineering and laying out work?
5. Can you select the proper tools?
6. Do you know how to time the delivery of materials?
7. Do you know how essential it is to have complete control of your job?

Almost any contractor confronted with the above set of questions will answer yes to all of them, yet there are large numbers that could greatly improve their methods. They are slaves of habit and insist on following the same pattern year in and year out. Here we are not so much interested in contractors in general as we are in the essentials of good management.

SELECTING MECHANICS

A contractor must know electrical-construction work well and understand mechanics in order to be able to man a job properly. Requirements of work and the ability of mechanics both vary greatly. Just because a man has a union card or calls himself an electrician it does not follow that he is well trained and proficient in all lines of electrical-construction work. His experience may be in a limited field.

The mechanic who has spent his time as a residential wireman will require time to get adjusted to factory or office-building work. On the other hand, industrial and commercial wiremen are not generally found to be good residential men.

Consideration must also be given to the selection of foremen. A man may be a good foreman for certain lines of work but a failure when trying to direct others. He may be able to direct the work of three or four mechanics well but unable to control a large crew.

At times a foreman is required to have more or less contact with the owners. In that case the tact and diplomacy of the man must be considered.

The contractor not only must have a faculty for selecting new men but must be able to recognize the ability of his regular mechanics as well. He must be familiar with his men in order to assign them advantageously. This applies to both new and existing projects.

The contractor who is able to recognize the ability of his mechanics can often build up a good crew out of what at first appeared to be just a gang of second-rate mechanics. Men must be assigned to work where they fit and must be paired off properly.

A great number of mechanics are good workmen but poor leaders. Others like to take the lead. Again we have men whose dispositions are such that they cannot be placed with the average run of mechanics. The contractor must take into consideration all the characteristics of his men and try to work out the best combinations.

Some may think that the way to select men is by hiring and firing until a good crew is built up. Such practice is costly. Money is lost while the poor mechanic is on the job, and it costs money to get the replacement mechanic started. Hiring and firing also tend to demoralize the crew of better workmen.

Every effort must be made to select the right men the first time. If the better mechanics are not available, men should be selected who are apt, willing to learn, and can readily adapt themselves to the work. The experienced contractor can pretty well guess the characteristics of new men when interviewing them.

For the benefit of the student reader, a word about interviewing mechanics is in order. It is usually a case of asking a man what kind of work he has been doing, the names of previous employers, and how long he has worked for each. The contractor may or may not call up previous places of employment. Usually he relies on his own judgment as to the honesty and ability of the man.

It is not found necessary to have mechanics fill out long application forms. Chances are that the majority of mechanics, if given forms, would throw them in the wastebasket. In the first place they are not accustomed to filling out applications for employment, and in the second place, they would think the contractor was too fussy to work for.

DIRECTING THE MEN

Once a contractor has built up a good organization, he must keep it intact. To do this he must treat his men and direct his work in a manner that will make him a desirable employer.

Along with the support of the job goes the normal courtesy toward the men. Each man is an individual personality and must be regarded accordingly. This must be taken into account when pairing off men and assigning work.

Some men have better temperaments than others for supervising mechanics. One should be able to control his work rigidly without having the men take exception to his methods. A large majority of the electricians are high-grade men and like to be made to feel that they are working with the contractor rather than for him.

THE NEED OF JOB DRAWINGS

Being able to prepare installation plans is one thing, and being able to recognize the needs of a particular job is quite another. From the standpoint of an architect or engineer, a job may be well laid out, but from the standpoint of installation, there will be many details to be supplied. A contractor must be able to study the plans and foresee the construction needs.

In a previous chapter installation drawings were given for a one-story factory. They provided the layout but would be augmented with shop drawings and details showing knockout templates for cabinets and boxes, supporting frames, and in some cases, detailed dimensions of equipment locations.

On a multistory building, the feeder system alone may require many special drawings. Dimensioned location of conduits must be shown for both horizontal and vertical runs, hangers and shaft supports detailed, templates for cabinets and boxes given, and many other details provided.

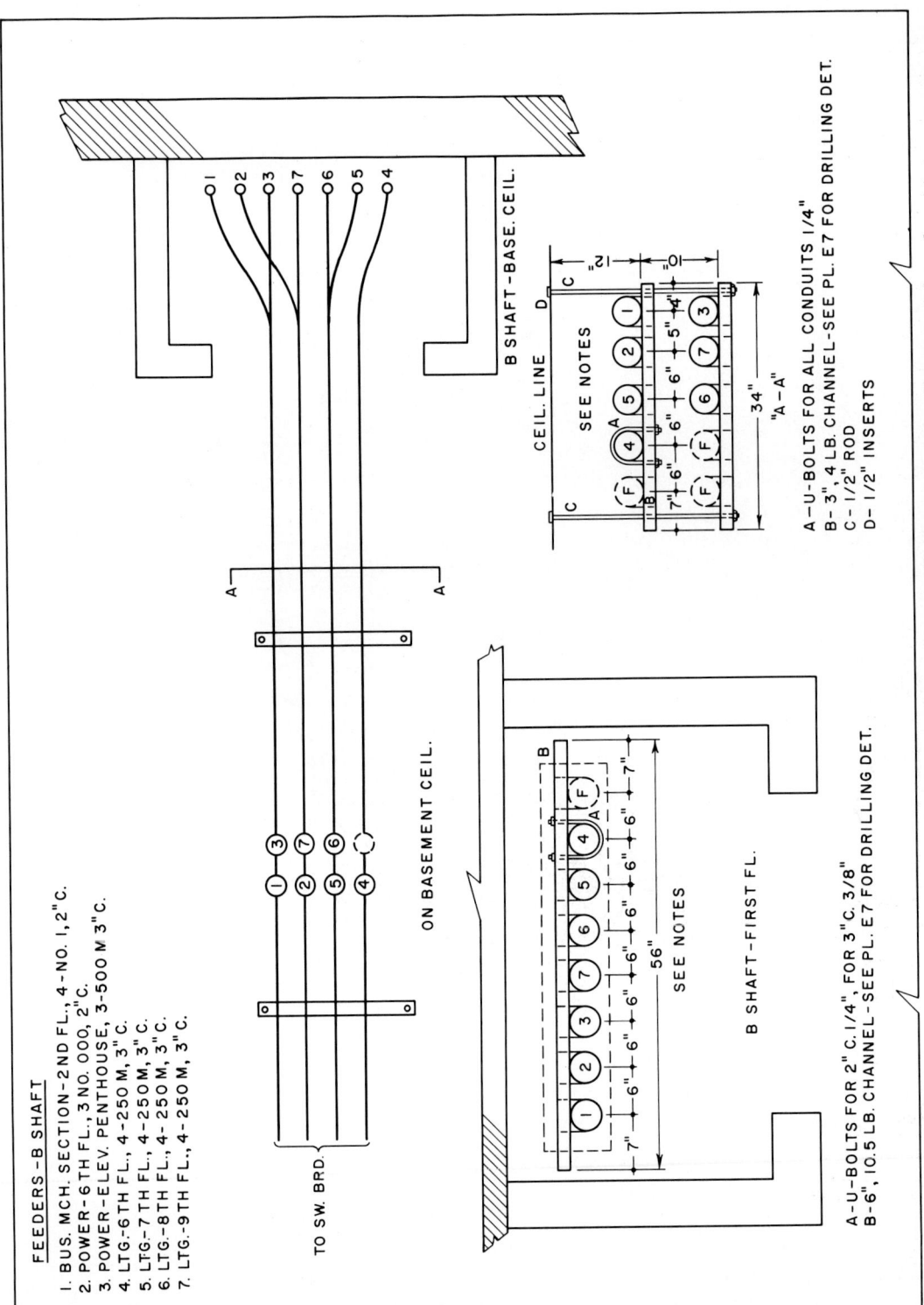

FIG. 9-1. Much planning is required for feeders entering crowded pipe shafts. Conduits must be arranged so that (*a*) a neat entrance to the shaft can be made, (*b*) they will be accessible at the right floors, (*c*) they can be properly supported, (*d*) pull and cable-support boxes can be placed where required.

The contractor not only must be able to foresee the needs of the installation but must be able to foresee them in time to have the drawings in the shop or on the job when needed. How often many of us have stayed at the office late, working on drawings for the job just opening up, so they would be out the next morning. It costs money to have construction drawings late.

Just why some contractors are so reluctant about supplying plans for their work is hard to understand. They throw a set of plans at the foreman and tell him to do the job. This is a costly practice and responsible for the expansion of many a contracting business being limited. They cannot compete for the bigger jobs and often close the doors to the offices of good architects and engineers.

Most architects and engineers cannot afford the time required to guide the contractor who cannot foresee his needs for job plans. If they have him on one job, they try to avoid the same mistake twice. He may get back a second time because he is the only contractor that has a price low enough to fit the budget.

ENGINEERING ABILITY

We have already stressed the necessity of contractors having engineering and design ability. Unfortunately, not all contractors recognize their deficiency in this respect.

Some contractors appreciate job needs and their lack of ability as well. For special jobs, they hire an outside engineer. Thereby they are able to enjoy the benefits of good electrical-construction engineering without having the burden of an engineer on the payroll at all times.

TOOLS

The importance of adequate tools has already been pointed out. The question to be settled by the contractor is, What constitutes adequate tools for the particular job?

Tools are expensive, and it costs money to get them to and from a job. Hence the contractor must know just what tools are needed to get best results.

A contractor does not always measure the utility of tools in the apparent dollar-and-cents value to the work. At times he may send out tools that will be used very little and could be dispensed with on the particular job, the purpose being to keep up the morale of the men.

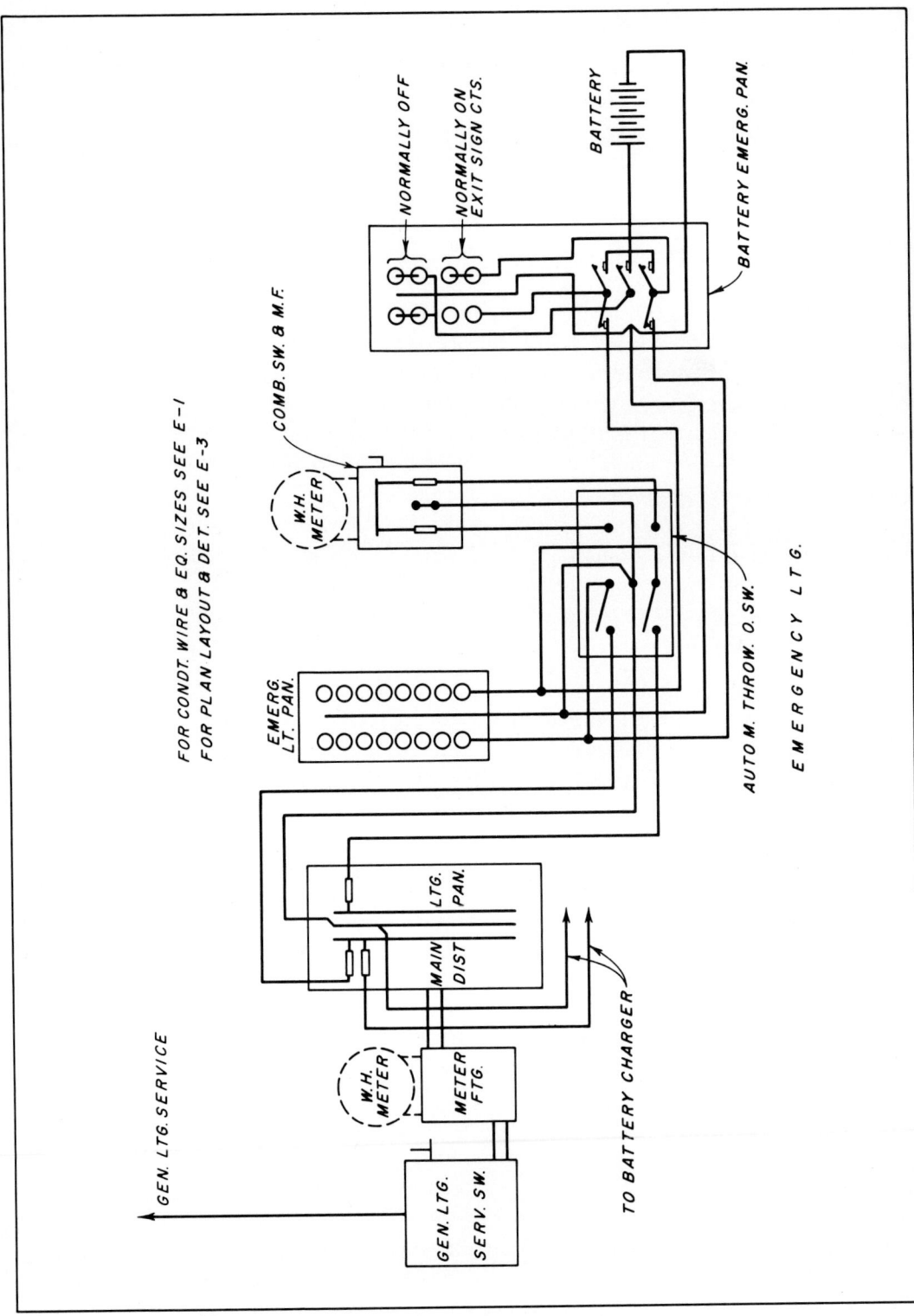

FIG. 9-2. Estimators are often obliged to make drawings hastily and send them directly to the field. A drawing may not be neat, but if it clearly shows what is to be done, it is valuable.

They are impressed with the importance of the work if they have the best tool for every purpose.

MATERIAL DELIVERIES

In *Electrical Estimating* (McGraw-Hill Book Company, Inc.), the author made the following statement: *"The right materials in the right quantities at the right time,* constitutes one of the cardinal rules of good construction practice."

To get the best results, one not only must appreciate the importance of this rule but must have a faculty for foreseeing the needs of the job as well.

The effect of delivering the wrong materials or delayed deliveries is practically the same. In either case the work is interrupted and men have to be shifted, be laid off, or be idle on the job. Such interruptions are costly to the work both directly and indirectly—directly because time is lost and indirectly because the interest of the mechanics is dulled.

The quantities must also be right. Shortage in quantities, if not discovered in time, will have the same effect as delayed deliveries. Excess materials create excess expense. There are cartage on the material out and back and the cost of handling it five or six times. Often material returned from jobs is too shopworn to be used on new installations.

In case of doubt, it is generally found better to ship a little long on materials. For the cost involved, one cannot take a chance on a job shortage.

Material deliveries must be timely. Materials delivered long before needed involve expense. The cost of storing and protecting can be considerable.

Designing hangers and pull boxes, getting templates in the shop, and deciding on special equipment are all part of the material delivery program. Too much emphasis cannot be put on this phase of electrical construction.

A contractor can never relax his hold on any phase of the work. Just as soon as he does, some part of the job starts to slip and is out of step with the rest of the project.

False economy may be responsible for a contractor's losing control of his work. Figure 9-3 shows the effect of management on the over-all cost of a project. Contractor *A* spends more money than contractor *B* to support his work, but the over-all cost of the completed installation is less and the duration time shortened. Figure 9-3 is for a large

ELECTRICAL CONSTRUCTION

Effect of Labor Management on Total Job Costs for a Project Requiring 50 Electricians

Item	Contractor A Number on job	Contractor A Average salary	Contractor A Extension	Contractor B Number on job	Contractor B Average salary	Contractor B Extension
1. Direct job costs per week:						
a. Personnel and salaries:						
Field engineer............	1	$250	$ 250	1	$160	$160
Draftsmen..............	2	120	240			
Clerk (1), timekeeper (1)	2	80	160	2	70	140
Accountants............	2	130	260	2	120	240
Stenographer...........	1	70	70	1	60	60
File clerk and telephone attendant.............	1	60	60	1	60	60
Material superintendent and tool mechanic.............	1	140	140	1	80	80
Assistant to material superintendent........	1	90	90			
Utility boy.............	1	50	50	1	50	50
Totals...............			$1,320			$790
b. Tools and miscellaneous direct job costs:.....						
Tools—construction and depreciation.....	..	...	$ 240	..	...	$100
Cartage—tools.........	..	...	6	..	...	4
Travel expenses........	..	...	14	..	...	6
Job stationery and Bl. P.	..	...	8	..	...	4
Job office equipment...	..	...	30	..	...	20
Telephone and telegraph............	..	...	30	..	...	10
Association dues.......	..	...	7	..	...	
Totals...............	..	...	$ 335	..	...	$144

Fig. 9-3. Operating expense must be sufficient to meet job requirements. Otherwise, labor costs will be excessive.

ELECTRICAL CONSTRUCTION
Effect of Labor Management on Total Job Costs for a Project Requiring 50 Electricians

Item	Contractor A Number on job	Contractor A Average salary	Contractor A Extension	Contractor B Number on job	Contractor B Average salary	Contractor B Extension
2. General overhead and administrative expenses per week—pro rata						
Administrative salaries........	½ x 500		$250	½ x $300		$200
Accountant..........	..	...	30	..	...	20
Stenographer.........	..	...	20	..	...	10
Telephone operator....	..	...	20	..	...	10
Rent (including heat)..	..	...	60	..	...	40
Telephone...........	..	...	12	..	...	8
Office equipment and supplies...........	..	...	24	..	...	10
Light...............	..	...	4	..	...	3
Taxes, legal, and advertising.........	..	...	7	..	...	2
Miscellaneous expenses.	..	...	5	..	...	4
Totals............	..	...	$432	..	...	$307

Comparisons

Item	Contractor A	Contractor B
1. Direct job costs—a......................	$1,320	$ 790
Direct job costs—b......................	335	144
2. General overhead and administrative expenses.	432	307
Totals................................	$2,087	$1,241
Duration of project........................	32 weeks	40 weeks
Cost of complete installation (materials not included):		
Direct labor—normal.....................	$256,000	$256,000
Excess labor............................		51,200
Operating costs.........................	66,784	49,640
Totals................................	$322,784	$356,840

Operating costs for A 65% greater than for B
Cost of complete installation, 10% greater for B than for A

FIG. 9-3. (*Continued*)

project; however, small installations show similar effects of management.

Although the values in Fig. 9-3 are for two different contractors, they could just as well represent two experiences of one contractor. *A* could represent work done under normal conditions, and *B* work done when contracts exceeded the capacity of the organization. Every organization has its limits, and any contracts beyond that limit must suffer from lack of good management and pay a toll of excess labor.

CHAPTER 10

*Are You Familiar with
Operating Costs?*

Do you know just what is meant by the term *operating costs,* and do you know how to figure *overhead?* It is as essential for a contractor to understand operating costs as it is for him to be able to prepare accurate material and labor costs.

The subject of operating costs is too extensive to be covered in a single chapter. In this chapter we shall get acquainted with the subject and learn how futile it is to try to base the costs of our own business operations on values established by others. The latter will be accomplished by studying some early surveys.

A man entering the electrical contracting business should be able to elaborate intelligently on answers to the following questions:

1. Do you know that the only place a contractor can learn about his overhead is in his own office?
2. Do you know the difference between direct job costs and overhead?
3. Do you know that overhead markups must be varied to fit the particular job?
4. Do you know that the cost of supplying labor is much greater than the cost of supplying material?
5. Do you know that countless numbers of contractors never realize the amount of profit that is indicated as such on their estimate sheets?

In previous chapters, markups were used in examples and operating costs discussed. Any duplication here will be of items that bear repetition.

In the study of operating costs, the following definitions must be thoroughly understood:

Base Cost of Material. The purchase price of the material used in the installation.

Base Cost of Labor. The installation payroll.

Base Cost of the Job. Combined base cost of material and labor.

Direct Job Costs. Costs of operating, such as estimating, engineering, supervision, tools, etc., which can be identified with a particular job.

Overhead. Items of expense, such as rent, heat, light, bookkeeping, telephone, and administrative, which cannot be identified with any particular job.

Operating Costs. All costs over and above base costs required to deliver a complete job.

Complete Job Cost. Base costs plus operating costs.

Profit. A permanent gain which can be taken out of the business. There are many other definitions (see Definitions) which we must understand, but the foregoing will serve for our explanations here.

Direct Job Costs

Operating costs are generally divided into two groups: direct job costs and overhead. Too often, contractors set up a group of all expenses over and above the base cost and call it overhead.

Direct job costs must be kept out of overhead if one is to have accurate estimating and billing costs. Much money has been lost by contractors because they did not know how to bill for tools, engineering, and other job costs.

Items of job cost vary greatly with the type and size of the project. Engineering is among those most likely to show a wide variation. Some work has well-laid-out plans, and the contractor's engineering is limited to shop and construction drawings. Other work comes to the contractor in more or less of a skeleton form, and he has to figure feeders, design panels and switchboards, figure lighting intensities, and provide all details of the installation.

The cost of tools varies with the size and type of project. It costs more per dollar of labor to supply tools for a five-man job than it does for a fifty-man job. However, tools are usually billed as a percentage of the payroll without too much consideration being given to the size and type of project. In addition, special tools are charged directly to the work.

In some cases all tools are charged directly to the job the same as material. At the close of the job, they are either turned over to the customer or bought by the contractor at some price agreed upon.

Overhead

In a later chapter, overhead will be studied at length. We shall then see why it is so essential for each contractor to study his own business instead of trying to rely on others to get information about overhead costs. A following study of operating-cost surveys will also serve to emphasize this point.

Profit

Profit has been defined as a permanent gain which can be taken out of the business. Some contractors take exception to this. They say that gains put back into the business are profit. Anything going into the business is subject to the hazards of the business. Take tools, for example.

Tools are about the most stable thing a contractor buys, yet they are subject to the hazards of the business. A contractor may buy $1,000 worth of tools, and then a slump comes along. He cannot use them, and no one wants to buy them. If the slump in business is prolonged, the tools have to be stored and kept in shape. When business opens

```
                    PROFIT?
     THE SO CALLED PROFIT DISAPPEARS.
                     WHY?
     BECAUSE IT IS —
        1— USED TO PAY DIRECT JOB COSTS
        2— CONSUMED BY GEN. OH. & ADMIN. COSTS TO-
           A— MAKE UP DEFICIT IN ADMIN. SALARIES
           B— PAY BONUSES, TO EMPLOYEES, WHICH ARE
              GIVEN IN LIEU OF SALARY INCREASES
           C— PAY INTEREST ON INVESTMENT
           D— REPLACE EQUIPMENT
           E— CARRY BUSINESS THROUGH DULL PERIODS
        3— RETURNED TO THE BUSINESS TO COVER
           UNFORESEEN EXPENSES AND POTENTIAL
           HAZARDS
```

FIG. 10-1. A business reminder.

up again, he may find that he has spent all that money on tools that have become obsolete.

Profit Disappears

Many contractors consistently add large percentages to their estimated costs for profit. When the books are closed at the end of the year, there are no such percentages shown for profit, the reason being that the contractors were just as consistently underestimating operating costs.

As we shall see later, establishing overhead markups for individual contractors and individual projects is a major task. It is possible to get reliable data that will greatly assist the contractor, but he must do considerable studying for himself.

A study of early surveys will enable us to understand why data on overhead supplied by others may be misleading.

A STUDY OF SURVEYS

Not many years back, a common question among contractors was, "How much is overhead in the electrical contracting business?"

In 1939, the author had occasion to give a talk in Philadelphia before a group of electrical contractors. Of the 150 or more present, most of them were asking, "How much is overhead?"

Back in their home towns, hundreds of other contractors were asking the same question. Prior to that time, nothing had been supplied to show them the necessity of relying on studies of their own business for the answer.

Most contractors thought that surveys would answer their questions about overhead. Prior to 1939 some surveys had been sponsored by contractors' organizations and the findings given to their members. The members still were not satisfied. A study of Fig. 10-2 will explain this.

Figure 10-2

To one familiar with operating-cost studies, it is evident that the persons conducting the surveys reported in Fig. 10-2 were not ready for the task. The questions were evidently too few and too vague. It is also evident that most of the answers were not enlightening.

There are seven divisions of expenses shown in Fig. 10-2. Figure 10-3 shows a total of 38 items. Of the 38, 19 are direct-job-cost items and 19 are items of overhead expense. In studying overhead it is well

Are You Familiar with Operating Costs? 85

to study the direct job costs at the same time because many expenses must be prorated. This will become obvious as we go along.

Let us study the detailed items in Fig. 10-2. The first item is administrative salaries. In some questionnaires, this item was listed as

Reports from Early Surveys of Overhead Costs						
Data Shown in Percentages of the Cost of Material and Labor						
Item	Dun and Bradstreet		A national trade association		Recc. by an accounting firm	Individual contractor's group
	1934	1935	1930	1937	1928	1938
Administrative salaries...........	12.0	13.5	9.85	↑	↑	6.50
Salary of office employees.........	14.5	17.4	8.85			11.22
Rent...............	3.48	2.5	1.2			1.14
Advertising.........	1.06	0.8	0.1	29.0	30.0	0.20
Light and heat.......	1.29	0.9	0.56			0.12
Taxes..............	0.93	0.7	0.74			
All other expenses....	8.80	10.3	16.8	↓	↓	10.67
Totals..........	42.06	46.10	38.10	29.00	30.00	29.85

The above values give some idea of overhead costs in the electrical contracting business but provide nothing that can be used for the individual contractor or the individual project. A careful study should not require any such allowances for "All other expenses."

Fig. 10-2. Results of early surveys indicating the lack of complete accounting records.

"Owner's and officers' salaries." Such salaries are not necessarily items of overhead expense. The owner or officers may engineer, superintend, or provide other services that should be charged off as direct job expense. If we look back at Fig. 2-1, we shall find that only 20% of the proprietor's time is charged to administrative work. Superintending, engineering, and a portion of the estimating belong to the job costs.

LABOR OVERHEAD VS. MATERIAL OVERHEAD
FOR ELECTRICAL CONTRACTING

APPROX. PERCENTAGES FOR BUSINESS WITH AN ANNUAL VOLUME (BASE JOB COSTS) OF 500,000 DOLLARS BASED ON SURVEYS OF CONTRACTING FOR A MIXED CLASS OF WORK WITH: (1) AN APPROX. JOB COST RANGE OF 100. TO 80,000. DOLLARS, AND (2) A CONTRACT RATIO OF "60-40"—60% OF THE BASE JOB COSTS IS FOR MATERIAL AND 40% FOR LABOR.

EXHIBIT-A (FOR REFERENCE ONLY)

APPROX. DIVISION OF DIRECT JOB COSTS WHICH SHOWS THAT, ON A DOLLAR BASIS, THE COST OF THE SUPPLY AND MANAGEMENT OF LABOR IS FAR GREATER THAN THE COST OF SUPPLYING MATERIAL.

NOTE—ITEMS OF DIRECT JOB COSTS SHOULD NOT BE CONFUSED WITH OVERHEAD.

ITEMS OF DIRECT JOB COST	MAT (60%)	LAB (40%)	JOB (100%)	REMARKS
ESTIMATING	1.05	1.35	1.17	SOME PROJECTS ARE ESTIMATED &ENG SEVERAL TIMES EXP. ON SINGLE PROJ. DOES NOT REFLECT COSTS FOR ADJ.
ENGINEERING & DRAFTING	0.80	2.88	1.63	BETWEEN JOBS, VACATIONS, ETC.
BLUE PRINTING	—	0.1	0.04	COST OF MISSIONARY WORK IS INCL IN GEN OH.
FIELD SHOP & OFFICE BLDGS	0.14	0.19	0.16	
FIELD TEL	0.03	0.03	0.03	FOR SPECIAL JOBS-SHOWN AS A PRORATA EXP. OF TOT. VOL.
WIRING & CURRENT—FIELD SHOP & OFFICE	0.05	0.10	0.07	
TOOLS-CONSUMED & DEPRECIATED	—	3.50	1.40	
SELECTING & PURCHASING MATERIAL	0.75	—	0.45	
FOLLOW UP & COORD DEL	0.50	—	0.30	TOOLS
CARTAGE & SPECIAL DEL	0.10	0.10	0.10	
SUPERVISION	—	2.60	1.04	
TRAVEL EXP—OFFICE TO JOB	0.10	0.21	0.09	
TIME KEEPER	—	0.50	0.20	REQ. ON THE JOB FOR LARGE CREWS
INSURANCES & EMPLOYEES BENEFITS	14.00		5.60	CHICAGO RATES
INSPECTION (CITY)	0.70	0.80	0.74	BASED ON CHICAGO RATES
INTEREST ON PAYROLL	—	0.50	0.20	CONTRACTOR IS ENTITLED INT. ON MONEY WITHELD
ASSOCIATION DUES	0.20	0.70	0.40	CHICAGO RATES
PRORATA CHARGES	—	0.05	0.02	FOR TEMP LT & PR, BROKEN GLASS, ETC.
RESERVE FOR CONTING & GUARANTEE	0.10	0.50	0.26	
TOTALS	4.43	28.11	13.90	

①—IN CHECKING OVERHEAD 14.11 (28.11−14.00) IS USED INS (14%) IS NOT REQ TO CARRY THE SAME MARKUP AS LAB

—THE CHIEF FUNCTION OF ELECTRICAL CONTRACTING IS THE SUPPLY & MANAGEMENT OF LABOR—

EXHIBIT-B
APPROX. DIVISON OF GEN. OVERHEAD & ADMIN. EXPENSE
ITEMS OF OPERATING COSTS NOT IDENTIFIED WITH ANY PARTICULAR PROJECT

ITEMS OF OVERHEAD EXP.	MAT	LABOR	THE JOB
ADMINISTRATIVE SALARIES	2.35	8.35	4.75
ENGINEERING & EST. (MISSIONARY WORK)	0.40	1.40	0.80
BOOKKEEPING—GEN.	0.92	1.40	1.11
BOOKKEEPING—SPECIAL TAXES & INS.	0.20	1.57	0.75
STENO. & TEL. OPER.	0.25	0.88	0.50
STORE RM. ATTENDANT & SHOP MECH.	0.35	1.23	0.70
UTILITY BOY	0.18	0.65	0.37
RENT—OFFICE (HEAT INCL.)	0.35	1.23	0.70
RENT—STORE ROOM (HEAT INCL.)	0.10	0.35	0.20
LIGHT	0.06	0.21	0.12
TELEPHONE	0.25	0.50	0.35
OFFICE EQUIP. & FURNITURE	0.12	0.43	0.24
STATIONERY, EST. FORMS & MISC. SUPPLS.	0.23	0.40	0.30
POSTAGE	0.10	0.20	0.14
TAXES, LICENSES & LEGAL EXP.	0.12	0.43	0.24
ADVERTISING	0.12	0.43	0.25
INSURANCE ON EQUIP. & MISCL. EXP.	0.33	1.13	0.65
RESEARCH & TIME STUDIES	—	0.40	0.16
AUTOS, TRAVEL & MISC. PROMT'L. EXP	0.33	1.13	0.65
ADJ. FACTOR (7 NORMAL & 3 BAD YRS IN TEN / ORG MUST BE MAINT IN OFF YRS)	0.20	1.57	0.75
TOTALS	6.96	23.89	13.73

ELECTRICAL CONTRACTORS' ASS'N OF CITY OF CHICAGO
RWA

Fig. 10-3. A comparison of material and labor operating costs which shows that costs of supplying labor are much greater than those for material.

The second item in Fig. 10-2 is "Salaries of office employees." Estimators and engineers are commonly considered to be office employees, but a large portion of their work is chargeable directly to some job. In every survey, the amount given for salaries was out of proportion to the totals.

The items of rent, heat, light, and telephone are all overhead expenses because they cannot be identified with any particular job.

All Other Expenses. The item of "All other expenses" in Fig. 10-2 must be given special attention because it is all out of proportion. A good study of overhead costs should not have over 3 or 4% for unidentified items. In the surveys this item ranges from 8.8 to 16.8%. In the third survey listed, the total for all expenses was 38.10% and the amount listed for "All other expenses" was 16.8%. The figure 16.8 is approximately 44% of 38.10.

Not all contractors agree on where job costs leave off and overhead begins. The only way to get accurate figures is to have all operating costs carefully listed. From the complete listing, the overhead expense can be separated from the job costs.

STUDIES SERVE AS GUIDES

At best, the most reliable studies made by others can serve only as guides for the individual contractor. From a study of the listings and values given in Fig. 10-4, one can learn much about analyzing operating costs. He can also learn much about the relative cost of supplying material and labor, the relative cost of various types of work, the items of expense involved, and numerous other pertinent facts about operating costs.

The reader should not be surprised to find studies in this book or elsewhere with values that do not exactly correspond to those given in Fig. 10-4. Each illustration represents a separate study, and variations in results serve only to augment the statement that there are no sets of operating costs that are applicable to all contractors and all jobs.

DIVISION OF OPERATING COSTS FOR ELECTRICAL INSTALLATIONS
SHOWN IN PERCENTAGE OF COSTS OF MATERIAL AND LABOR

ITEMS OF EXPENSE	INDUSTRIAL WORK											
NOTE:- THE FIGURES IN THESE TABLES ARE BASED ON RELIABLE SURVEYS. SEE LIST BELOW. ①	LABOR & MATERIAL FUR. BY CONTR.											
	JOB REQ. 5 TO 25 ELECT.						JOB REQ. 25 TO 50 ELECT.					
	ENG. BY OTHERS			ENG. BY CONTR.			ENG. BY OTHERS			ENG. BY CONTR.		
	APPROX. MIN.	APPROX. MAX.	SEE NOTE 3	APPROX. MIN.	APPROX. MAX.	SEE NOTE 3	APPROX. MIN.	APPROX. MAX.	SEE NOTE 3	APPROX. MIN.	APPROX. MAX.	SEE NOTE 3
	1	2	3	4	5	6	7	8	9	10	11	12
I COST OF HANDLING MATERIALS ②												
ENGINEERING & LAY OUT	—	—		5.00	7.00		—	—		4.50	6.50	
SELECTING & PURCHASING	1.00	1.50		0.75	1.00		0.75	1.25		0.50	0.75	
FOLLOW UP & COORD. DELIVERY	0.50	1.00		0.30	0.50		0.40	0.80		0.25	0.40	
CARTAGE & SPECIAL DELIVERY	—	0.25		—	0.25		—	0.15		—	0.15	
STORAGE FACILITIES (FIELD)	0.04	0.50		0.04	0.50		0.02	0.40		0.02	0.40	
GEN. OVERHEAD & ADMIN. EXP.	5.65	10.58		5.65	10.60		4.55	8.50		4.80	8.80	
TOTALS	7.19	13.83		11.74	19.85		5.72	11.10		10.07	17.00	
II COST OF SUPPLYING LABOR												
SUPERVISION	2.60	3.20		2.34	2.88		2.44	2.88		2.20	2.60	
TOOLS—CONSUMED & DEPR.	3.50	5.70		3.50	5.70		3.00	4.00		3.00	4.00	
FIELD SHOP TOOLS	—	0.12		—	0.12		—	0.10		—	0.10	
CARTAGE (TOOLS)	0.10	0.20		0.10	0.20		0.10	0.15		0.10	0.15	
FIELD OFFICE & SHOP BLDG.	—	0.50		—	0.50		—	0.30		—	0.30	
ENG., LAY OUT, & STUDY TIME	—	—		5.00	7.00		—	—		5.00	7.00	
FIELD ENGINEERING	3.20	6.40		2.40	4.80		3.20	6.40		2.40	4.80	
FIELD TELEPHONE	—	0.14		—	0.14		—	0.14		—	0.14	
BLUE PRINTING & JOB STATIONARY	0.08	0.12		0.08	0.12		0.08	0.12		0.08	0.12	
TRAVEL EXP.—OFFICE TO JOB	0.15	0.35		0.15	0.35		0.10	0.25		0.10	0.25	
GEN. OVERHEAD & ADMIN. EXP.	14.86	21.20		14.16	20.00		12.00	17.00		11.30	16.00	
TOTALS	24.49	37.93		27.73	41.81		20.92	31.34		24.18	35.46	
LABOR ONLY JOBS—ADD EXP. OF CHKG. MAT. & COORD. DEL.—1.5 X I (MAT.) ④												
TOTALS—NOT INCL. INS.	24.49	37.93		27.73	41.81		20.92	31.34		24.18	35.46	
INSURANCES												
TOTALS												

ITEMS OF EXPENSE	INDUSTRIAL WORK						COMMERCIAL WORK					
	LABOR ONLY BY CONTR.						LAB. & MAT.			LABOR ONLY		
	5 TO 25 ELECT.			25 TO 50 ELECT.			5 TO 25 ELECT.			5 TO 25 ELECT.		
	ENG. BY OTHERS			ENG. BY OTHERS			ENG. BY OTHERS			ENG. BY OTHERS		
	APPROX. MIN.	APPROX. MAX.	SEE NOTE 3	APPROX. MIN.	APPROX. MAX.	SEE NOTE 3	APPROX. MIN.	APPROX. MAX.	SEE NOTE 3	APPROX. MIN.	APPROX. MAX.	SEE NOTE 3
	13	14	15	16	17	18	19	20	21	22	23	24
I COST OF HANDLING MATERIALS ②												
ENGINEERING & LAY OUT	—	—		—	—		—	—		—	—	
SELECTING & PURCHASING	CHKG. WITH & ASSTG. OWNER'S REP. 0.50	1.00		0.40	0.80		1.00	1.50		WORK WITH OWNER 0.50	1.00	
FOLLOW UP & COORD. DELIVERY	0.10	0.20		0.10	0.20		0.50	1.00		0.20	0.40	
CARTAGE & SPECIAL DELIVERY	—	—		—	—		0.50	1.50		—	—	
STORAGE FACILITIES (FIELD)	—	0.30		—	0.20		0.04	0.50		—	0.25	
GEN. OVERHEAD & ADMIN. EXP.	2.26	4.24		1.82	3.41		6.80	12.70		2.72	5.07	
TOTALS	2.86	5.74		2.32	4.61		8.84	17.20		3.42	6.72	
II COST OF SUPPLYING LABOR												
SUPERVISION	2.80	3.80		2.52	3.42		2.86	3.52		3.20	3.88	
TOOLS—CONSUMED & DEPR.	3.50	5.70		3.00	4.00		3.50	5.70		3.50	5.70	
FIELD SHOP TOOLS	—	0.12		—	0.10		—	0.12		—	0.12	
CARTAGE (TOOLS)	0.20	0.30		0.20	0.25		0.15	0.20		0.25	0.30	
FIELD OFFICE & SHOP BLDG.	—	0.50		—	0.30		—	0.50		—	0.50	
ENG., LAY OUT, & STUDY TIME	2.00	4.00		2.00	4.00		1.00	2.00		3.00	5.00	
FIELD ENGINEERING	2.40	4.80		2.40	4.80		3.20	6.40		4.00	8.00	
FIELD TELEPHONE	—	0.14		—	0.14		—	0.20		—	0.30	
BLUE PRINTING & JOB STATIONARY	0.30	0.50		0.30	0.50		0.10	0.15		0.20	0.30	
TRAVEL EXP.—OFFICE TO JOB	0.30	0.50		0.15	0.25		0.25	0.35		0.35	0.50	
GEN. OVERHEAD & ADMIN. EXP.	18.50	26.40		15.00	21.20		16.50	23.20		18.00	25.50	
TOTALS	30.00	46.86		25.57	38.96		27.56	42.34		32.50	50.10	
LABOR ONLY JOBS—ADD EXP. OF CHKG. MAT. & COORD. DEL.—1.5 X I (MAT.) ④	4.29	8.16		3.48	6.62					5.13	10.08	
TOTALS—NOT INCL. INS.	34.29	55.02		29.05	45.58		27.56	42.34		37.63	60.18	
INSURANCES												
TOTALS												

ITEMS OF EXPENSE	ALTERATIONS—COMMERCIAL								
	LAB. & MATERIAL BY CONTR.						LABOR ONLY		
	JOB REQ 5 TO 25 ELECT.						5 TO 25 ELECT.		
	ENG. BY OTHERS			ENG. BY CONTR.			ENG. BY OTHERS		
	APPROX. MIN.	APPROX. MAX.	SEE NOTE 3	APPROX. MIN.	APPROX. MAX.	SEE NOTE 3	APPROX. MIN.	APPROX. MAX.	SEE NOTE 3
	25	26	27	28	29	30	31	32	33
I COST OF HANDLING MATERIALS ②									
ENGINEERING & LAY OUT	—	—		6.00	10.00		—	—	
SELECTING & PURCHASING	1.00	2.00		0.75	1.50		WORK WITH OWNER 0.50	1.50	
FOLLOW UP & COORD. DELIVERY	0.50	1.50		0.50	1.00		0.50	0.75	
CARTAGE & SPECIAL DELIVERY	0.50	1.50		0.25	1.00		—	—	
STORAGE FACILITIES (FIELD)	0.04	0.10		0.04	0.10		—	—	
GEN. OVERHEAD & ADMIN. EXP.	8.50	15.80		8.50	15.80		3.41	6.43	
TOTALS	10.54	20.90		16.04	29.40		4.41	8.68	
II COST OF SUPPLYING LABOR									
SUPERVISION	4.00	5.00		3.60	4.50		5.00	6.25	
TOOLS—CONSUMED & DEPR.	3.00	4.60		3.00	4.60		3.00	4.60	
FIELD SHOP TOOLS	—	0.15		—	0.15		—	0.30	
CARTAGE (TOOLS)	0.35	0.45		0.30	0.40		0.40	0.50	
FIELD OFFICE & SHOP BLDG.	—	0.17		—	0.17		—	0.20	
ENG., LAY OUT, & STUDY TIME	1.00	2.00		6.00	8.00		3.00	5.00	
FIELD ENGINEERING	4.00	8.00		3.60	6.40		4.80	9.60	
FIELD TELEPHONE	—	0.40		—	0.40		—	.0.50	
BLUE PRINTING & JOB STATIONARY	0.40	1.00		0.40	0.60		0.50	1.00	
TRAVEL EXP.—OFFICE TO JOB	0.50	0.60		0.50	0.60		0.60	0.70	
GEN. OVERHEAD & ADMIN. EXP.	22.40	31.80		22.40	31.80		28.00	39.80	
TOTALS	35.65	54.17		39.80	57.62		45.30	68.45	
LABOR ONLY JOBS—ADD EXP. OF CHKG. MAT. & COORD. DEL.—1.5 X I (MAT.) ④							6.62	13.02	
TOTALS—NOT INCL. INS.	35.65	54.17		39.80	57.62		51.92	81.47	
INSURANCES									
TOTALS									

NOTES—

①—THE FIGURES IN THESE TABLES ARE BASED ON SURVEYS AND STUDIES MADE BY THE FOLLOWING:—
- DUN AND BRADSTREET — — — — 1934
- DUN AND BRADSTREET — — — — 1935
- NATL. ELECTL. CONTRS.' ASSN. — — — — 1937
- ELECTL. CONTRS.' ASSN. OF PITTSBURG — — 1928 (SURVEY BY A. H. BLASE CO., AUDITORS)
- ELECTL. CONTRS.' ASSN. OF CHICAGO — — — 1939
- ELECTL. CONTRS.' ASSN. OF CHICAGO — — — 1940

②—CONTRACTING (ELECTL.) IS A SERVICE TRADE WITH THE SUPPLYING & MANAGEMENT OF LABOR AS THE PRIMARY FUNCTION. MATERIAL IS INCIDENTAL AND REGARDLESS OF WHO PURCHASES IT, THE CONTRACTOR ASSUMES RESPONSIBILITY IN CONNECTION WITH THE PROPER SELECTION, DELIVERY, STORAGE, INSTALLATION, AND OPERATION. COMPENSATION MUST BE COMMENSURABLE TO SERVICES RENDERED.

③—FILL IN VACANT COL. WITH PERCENTAGES TO FIT THE PARTICULAR JOB BEING CONTEMPLATED.

④—IT IS ASSUMED THAT THE COST OF MATERIAL WILL BE 50% GREATER THAN THE COST OF LABOR, OR THAT 1% OF MAT. COST IS EQUAL TO 1.5% OF LAB. COST. FOR LABOR ONLY PROJECTS, 1.5 TIMES THE PERCENTAGE SHOWN UNDER MAT. HANDLING (ITEM I) MUST BE ADDED TO LABOR (ITEM II)

FIG. 10-4. A study of operating costs for various types of work.

CHAPTER 11

Do You Know How to Conduct Overhead Studies?

Do you know that in figuring the over-all cost of jobs, the errors made in estimating overhead costs are often greater than those made in estimating the cost of labor and material? Do you know how essential it is for a contractor to be able to establish his own experienced overhead costs? And do you know how to use studies of overhead costs prepared by others? Before we start our study of overhead, these three questions will be noted.

Research has disclosed that many offices, not familiar with operating costs, are using fatuous overhead markups. They base their markups on what someone else has told them is right, on what they have been using for years, or on what they think the traffic will bear. As a rule, such contractors have no listing of items for direct job expenses. All costs, over and above the base costs of labor and material, are listed as overhead. We have already noted the necessity of keeping direct job expenses out of overhead.

Contractors are more concerned about errors in labor than they are in overhead, yet on a 60/40 job, a 10% error on labor would be less than a 5% error in overhead markup on the complete job cost. On a $10,000, 60/40 ($6,000 material and $4,000 labor) job, 10% on labor would be $40 whereas 5% on the total cost would be $50.

In the previous chapter we had examples that stressed the importance of a contractor's knowing how to establish his own experienced overhead costs. Here we are interested in how he is to accomplish the task.

As we shall soon see, one who has access to and knows how to use reliable overhead studies prepared by others is at a distinct advantage over one who must start with nothing to guide him. Let us look at Figs. 11-1 to 11-3. The studies portrayed by these figures represent several years of research over a vast scope of electrical-construction work. Figures 10-3 and 10-4 and Fig. 2-3 are of a type that were numerous forerunners of Figs. 11-1 to 11-3.

From a preliminary study of Figs. 11-1 to 11-3, we learn how overhead costs may be expected to vary according to the size of the project and, in general, how the cost of supplying material will compare with the cost of supplying labor. In this chapter we shall learn what a great help such curves can be in establishing experienced overhead costs for the individual business. Let it be stated again that we are indeed fortunate to have such studies at our disposal.

PRELIMINARY JOB STUDIES

As a matter of general information, every office should have a tabulation of all contracts for each completed year, showing the following:

Estimated cost of material
Actual job cost of material
Estimated cost of labor
Actual job cost of labor
Material-labor ratio (the finished job)
Estimated direct job expense
Actual direct job expense
Estimated overhead
Estimated profit

From the study of such a tabulation, one learns a great deal about his business. He not only learns how his experienced material and labor costs compare with the estimated costs, but from the totals of direct job expenses and overhead, he learns whether or not his markups are in keeping with actual costs. Such studies do a great deal to get one started to thinking in the right direction and stimulate a desire to establish accurate operating costs.

Overlapping Work

In preparing for the study of any yearly costs, one will, in all probability, have overlapping contracts to reckon with. Such work will have

Fig. 11-1.

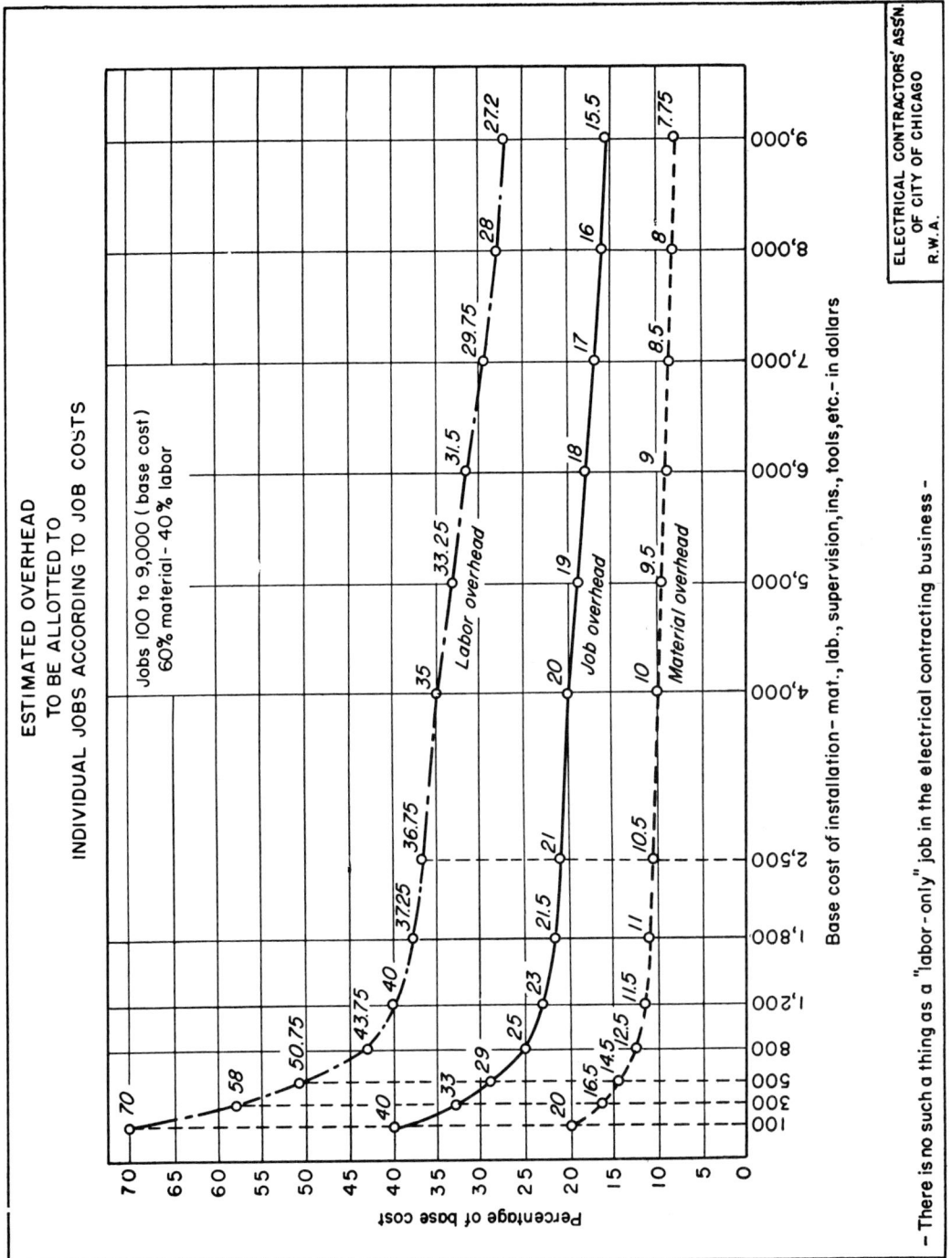

Fig. 11-2.

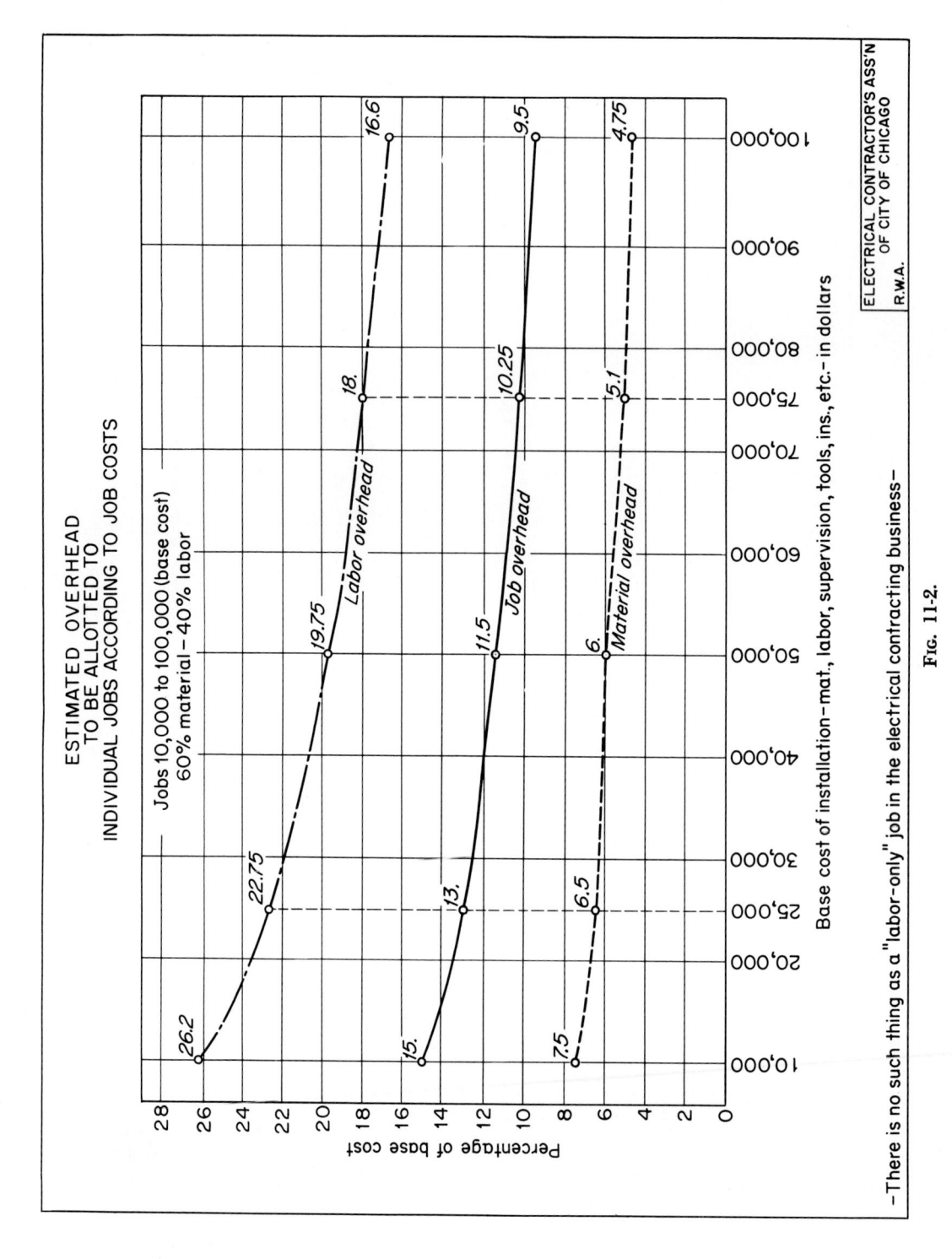

Fig. 11-3.

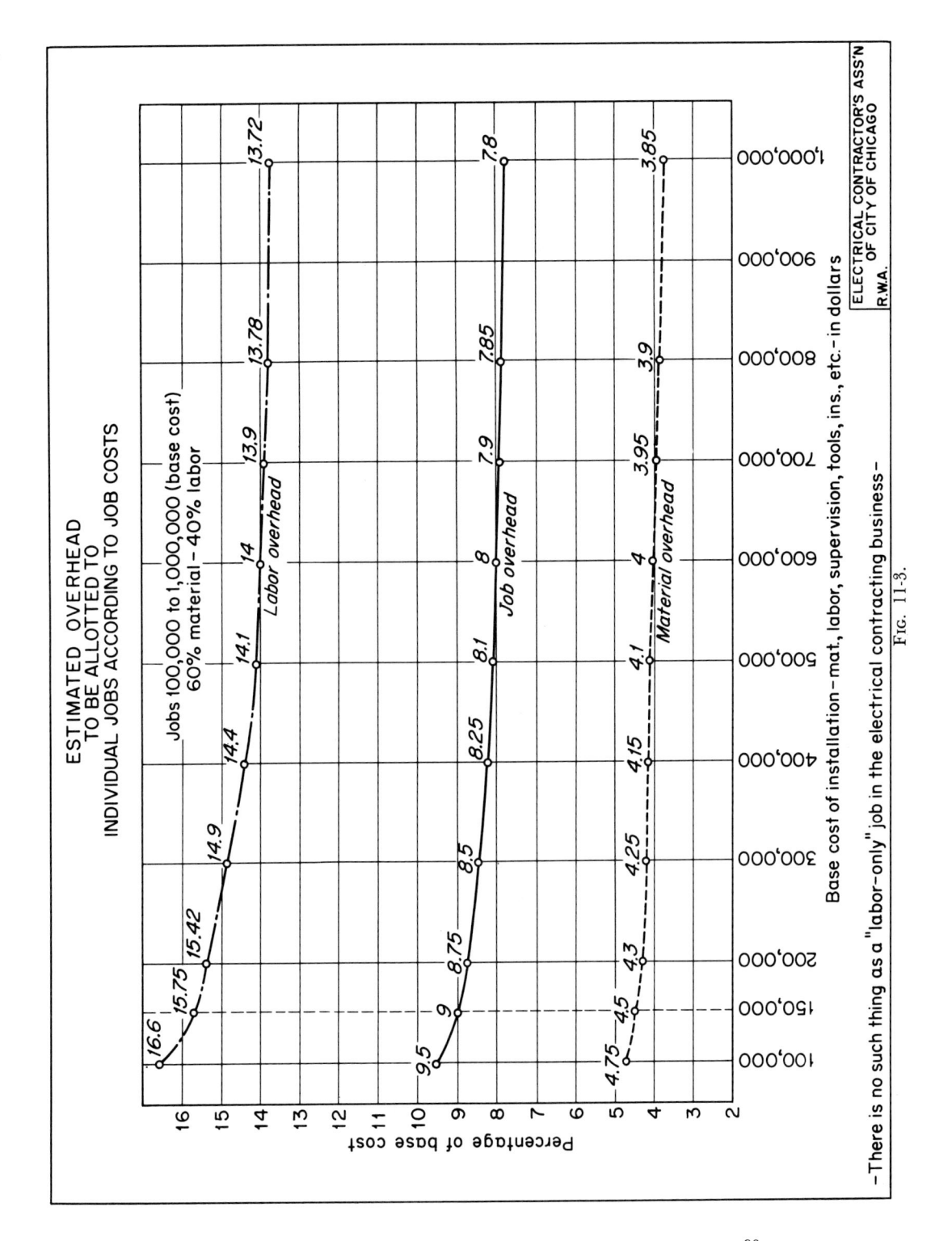

to be prorated. To prepare an accurate estimate of unfinished work is often difficult. However, we do not expect final figures to be 100% correct, and the errors resulting from such estimates should not be too serious.

Extra Orders

Extra orders may also prove difficult. Again one must establish costs as accurately as possible and be prepared to accept some slight errors. "Extras" are especially elusive to accurate accounting because they are interwoven into the main project and cannot be easily segregated. Often contractors include all extra orders with the main contract. This is all right if the estimates for the extras are accurate. With a large volume of extra work, figured too liberally, a combination with the main contract will produce figures that are misleading.

PRELIMINARY OVERHEAD STUDIES

Figure 11-4 shows a preliminary study of overhead costs prepared to give an idea of how experienced over-all costs compared with normal. Jobs of an approximate dollar volume are grouped, and round figures are used. The grouping is for $100, $300, $500, and so on. In the total, an amount must be included (plus or minus) to compensate for any excess or shortage produced by the grouping and use of round figures. In Fig. 11-4, $980 is added.

The percentage values used in the "Normal overhead" column, Fig. 11-4, were taken from the overhead curves shown in Figs. 11-1 and 11-2. Since this study is for overhead on the complete cost, the "Job overhead" curves were used. Values for all the groups, except those for the $10,000 and $18,000 contracts, were taken from Fig. 11-1. Values for the $10,000 and $18,000 contracts were taken from the "Job overhead" curve of Fig. 11-2.

The "Job curve," Fig. 11-1, shows 40% for $100, 29% for $500, 20% for $4,000, and so on, through the list. In Fig. 11-2, the "Job curve" shows 15% for $10,000 and 14% for $18,000. An estimated 25% was used for the $980 correction item.

The total resulting from the application of normal overhead costs is shown as $16,290. The contractor's experienced overhead was $15,660, which was found to be $800, or 5% less than the estimated normal overhead.

To get experienced values for estimating, the normal overhead values would have to be reduced 5%. Figure 11-1 shows a "Job overhead" value of 20% for a $4,000 base-cost contract with an M/L ratio of 60/40; 5% of 20 is 1. The contractor's estimating overhead for a similar $4,000 job would be 20 minus 1, or 19%.

<table>
<tr><td colspan="5" align="center">PRELIMINARY CHECK OF ANNUAL OVERHEAD (Over all)</td></tr>
<tr><td rowspan="2">No. of jobs</td><td colspan="2" align="center">Amount of contractors</td><td colspan="2" align="center">Estimate of normal overhead</td></tr>
<tr><td>Each</td><td>Exten.</td><td>%</td><td>$</td></tr>
<tr><td>10</td><td>Average 100</td><td>1,000</td><td>40</td><td>400</td></tr>
<tr><td>10</td><td>Average 300</td><td>3,000</td><td>33</td><td>990</td></tr>
<tr><td>20</td><td>Average 500</td><td>10,000</td><td>29</td><td>2,900</td></tr>
<tr><td>9</td><td>Average 800</td><td>7,200</td><td>25</td><td>1,800</td></tr>
<tr><td>10</td><td>Average 1,200</td><td>12,000</td><td>23</td><td>2,760</td></tr>
<tr><td>3</td><td>Average 2,500</td><td>7,500</td><td>21</td><td>1,575</td></tr>
<tr><td>2</td><td>Average 4,000</td><td>8,000</td><td>20</td><td>1,600</td></tr>
<tr><td>1</td><td>10,000</td><td>10,000</td><td>15</td><td>1,500</td></tr>
<tr><td>1</td><td>18,000</td><td>18,000</td><td>14</td><td>2,520</td></tr>
<tr><td colspan="2">Add to corrected total......</td><td>980</td><td>Estimated 25</td><td>245</td></tr>
<tr><td colspan="2">Totals..................</td><td>77,680</td><td></td><td>16,290</td></tr>
</table>

Estimated normal overhead.... $16,290
Experienced overhead......... 15,480
Below normal overhead....... $ 810

$$\frac{810}{16,290} = 0.0498 \text{ (use 5\%)}$$

FIG. 11-4.

A contractor who has been using overhead markups which check within 5% of normal would be wise to continue without change until he has a second-year check.

COMPLETE OVERHEAD STUDIES

Figure 11-5 illustrates a method of studying material and labor overheads for a complete listing of contracts. All major projects have a separate listing; only very small jobs and extra orders are grouped.

OVERHEAD STUDY FOR YEAR ENDING DEC. 31, 1956

NAME OF PROJECT	COMPLETE JOB COSTS			M/L Ratio	TOTAL COST OF JOB A+B+C	ESTIMATED NORMAL OVERHEAD							
	Mat. A	Labor B	Job Exp. C	D	E	Material % F	$ G	Labor % H	$ I	Job Expense % J	$ K	Totals %	$ G+I+K
Burley Machine Co.	1,580.	1,150.	110.	58/42	2,840.	10	158.	36	410.	18	20.	20.5	588.
Clark and Black	790.	750.	56.	52/48	1,596.	11	87.	38	285.	20	11.	24	383.
Brontons	440	320	30.	58/42	790.	13	57.	45	144.	25	8	26.5	209.
Oxford Building	6,200.	3,850.	340.	59/41	10,390.	7.5	445.	26	1,000.	16	54.	15	1,499.
Bell Academy	13,650.	6,340.	640.	68/32	20,630	7	956.	23	1,460.	15	96.	12	2,512.
Miscel. Jobs, 10 total	400.	450.	45	44/56	895.	20	80.	70	315.	45	20.	46	415.
Extra orders, 8 total	250.	330.	30.	43/57	610.	20	50.	70	231.	45	14.	48	295.
TOTALS	46,400	28,600	2,680.	62/38	77,680.		4,600.		11,100		540.	20.8	16,240.

Total Exper. Operating Costs ---- $18,160.
Direct Job Expences ---- 2,680.
Experienced Overhead ---- 15,480.

Estimated Normal Overhead ---- $16,240.
Experienced Overhead ---- 15,480.
OH. Below Normal ---- 760.

760 / 16,240 = .047 (use 5%)

FIG. 11-5.

As indicated by the broken lines, not all jobs were listed in the illustration. However, the totals are for all contracts. To have listed all contracts would have produced an ungainly sheet without improving the illustration.

In the illustration, Fig. 11-5, completed job costs are listed for material (Col. *A*), labor (Col. *B*), and direct job expenses (Col. *C*). The normal overhead for each of these divisions is listed in Cols. *F, H,* and *J,* respectively. In the same order, the estimated normal overhead in dollars is shown in Cols. *G, I,* and *K.* The estimated overhead for the complete job is shown under the heading "Totals."

Figures 11-1 and 11-2 were again used for establishing normal overheads. However, for Fig. 11-5, the "Material overhead" and "Labor overhead" curves were used.

As shown by Col. *D,* the contracts do not all adhere closely to the 60/40 M/L ratio. To avoid confusion, the values on the curves will be used without adjustment. Adjustment for contracts without a 60/40 M/L ratio will be discussed later. Using the curve values without adjustment will not introduce any great error because, as shown in the summary, Col. *D,* the over-all volume ratio is 62/38.

The first listing, Burley Machine Company, Fig. 11-5, has a total job cost (Col. *E*) of $2,840. Turning to Fig. 11-1 and looking at the $2,800 positions on the curves, we find the normal overhead values for material and labor. For material, the reading is slightly above 10%; 10% was used. For labor, the reading is approximatly 36.4%. To simplify calculating and checking, 36% was used (see Col. *H* in Fig. 11-5).

For the fourth listing, Oxford Building, Col. *E* shows a total cost of $10,390. Looking at the curves of Fig. 11-2, we see that the $10,000 readings can be used without introducing any great errors. On the $10,000 line, the material reading is 7.5% and the labor reading 26.2%; 7.5% was used for material and 26% was used for labor (see Cols. *F* and *H*).

The markups used for the job expenses (see Col. *J*) are selected according to the author's appraisal. It will be noted that in most cases an average of the material and labor markups has been used. For the Bell Academy, the labor overhead was 23% and the material overhead was 7%. For the job expense, an average of 15% was used.

In the summary, the total estimated normal overhead for the annual volume is shown as $16,240. Below the table, Fig. 11-5, is a list of calculations. The experienced operating cost is given as being $18,160, and the experienced overhead is obtained by subtracting the total direct job expense of $2,680 (see total Col. *C*), from that amount. The remainder is $15,480, which, as stated before, represents the contractor's experienced overhead expense for the year's volume.

Subtracting the experienced overhead of $15,480 from the estimated normal overhead of $16,240 leaves $760. The experienced overhead is $760 less than the estimated normal overhead, which is approximately 5%. Assuming that the study in Fig. 11-5 was based on 60/40 M/L contracts, this would indicate that the electrical contractor could use overhead markups 5% below normal.

The method of reduction would be the same as explained for Fig. 11-4, except there would be adjustments for two items, namely, material and labor. Referring to Fig. 11-1, we see that for a $4,000 project, the normal markups would be 10% for material and 35% for labor. Our study indicates that the contractor could use 5% less, or 9.5% for material and 33.25% for labor.

With all the hazards and variables in electrical-construction work, there are times when contractors do not regard a 1 or 2% markup too seriously. However, on the total volume 1 or 2% difference in markup would mean a great deal.

On a contract with a $200,000 payroll a change of 1% in the labor markup would make a difference of $2,000. Compared with the total cost of the contract, $2,000 would not be a great deal, but compared with the profit ordinarily enjoyed by electrical contractors, it would be an appreciable amount.

ADJUSTING FOR UNBALANCED RATIOS

As noted, the studies illustrated by Figs. 11-1 to 11-3 are based on contracts with an M/L ratio of 60/40. At the time the studies were made, a large majority of electrical contracts adhered closely to that division of costs for material and labor. There are times when contracts vary greatly from this ratio, and in studying operating costs we must know how to make adjustments.

For simple wording, we can call all contracts with a 60/40 M/L ratio balanced and those with other ratios unbalanced. An example of an unbalanced job is the Bell Academy, listed in Fig. 11-5. The M/L ratio as shown in Col. D is 68/32. We can use this contract to illustrate the method of adjustment.

Before proceeding, the reader should be informed that with a predominance of material cost as we have in our example, adjustments are not too complicated, but with a cost of labor greatly in excess of that for material, one has to be on his guard. The hazards of installation-only (material by others, installation only by contractor) must be considered.

Returning to our example, the Bell Academy, Fig. 11-5, we have a material cost of $13,650 and a labor cost of $6,340. We will first adjust the material MU (markup).

A 60/40 contract with a material cost of $13,650 would have a total cost of $22,750 ($13,650 material and $9,100 labor). Looking at Fig. 11-2, the material curve for a $22,700 project shows 6.5% as the proper overhead MU. In our example, Fig. 11-5, 7% was used. The adjustment would be 0.5%.

The labor must also be based on a 60/40 job. The cost of labor for the Bell Academy was $6,340. A 60/40 job with a labor cost of $6,340 would have a total cost of $15,850 ($9,510 material and $6,340 labor).

The curve for labor, Fig. 11-2, indicates that for a $15,850 contract, the labor overhead markup should be 24.5%. In our example Fig. 11-5, 23% was used.

We can apply the adjusted percentages and learn the extent of the change.

For material we have......	$13,650 @ 6.5%	$ 887
For labor we have.........	6,340 @ 24.5%	1,553
Total..............................		$2,440

In our original estimate, Fig. 11-5, we had $956 (Col. G) for material and $1,460 (Col. I) for labor. The total for labor and material (not including job costs) was $2,416. The difference between the adjusted and original estimate is $24. Not all jobs work out so closely.

MATERIAL SERVICE

During the study of installation-only (labor and installation services by contractor, material by owner) projects, we shall learn that the contractor has an expense resulting from services rendered in connection with materials furnished by others. The cost of such services belongs with direct job expenses and must not be included with overhead. Details of material service costs will be dealt with along with the study of installation-only projects. In a following paragraph we shall have an example of how to estimate such costs.

ESTIMATING DIRECT JOB EXPENSES

As a check on direct job expenses, it is good practice to have figures available showing the approximate percentage of the total material and labor costs represented by same. Such figures are arrived at by using known percentages as follows:

Estimate of over-all direct job expenses for a $10,000, 60/40 project:

Material direct job expense	$6,000 @ 2%	$120
Labor direct job expense	4,000 @ 6%	240
Total estimated direct job expense		$360

$$360 \div 10,000 = 0.036$$

Here, without insurances, we have an estimated 3.6% of the total cost of material and labor. In practice, 3.5% is often used.

Let us take a second example based on the following:

$10,000 installation (base cost labor and material)
$6,000 contract ($2,000 material, $4,000 labor)
$4,000 material by owner
Direct job expense, material by contractor, 2%
Direct job expense, material by owner (material service), 2%
Direct job expense, labor (not including insurance), 6%
Direct job expense, labor insurances, 14%

The estimate of over-all DJE (direct job expense) is as follows:

Material DJE	$2,000 @ 2%	$ 40
Material service	$4,000 @ 2% (estimated)	80 (material by owner)
Labor DJE—general	$4,000 @ 6%	240
Labor DJE—insurance	$4,000 @ 14%	560
Total estimated DJE		$920

Here the estimated direct job expense of $920 is 9.2% of the total installation cost. However, in terms of the $6,000 contract, it represents 15.3%.

For unbalanced contracts, separate percentages must be figured for each M/L ratio. In actual practice, one would more than likely take 3.5% of the estimated installation cost plus the insurance. For the above example he would have 3.5% of $10,000 plus 14% of $4,000.

VALUE OF LIMITED STUDIES

It is not expected that the majority of contractors will make such detailed studies of overhead as have been illustrated. However, there are others who have their own methods and go into more detail.

At first, one may find the study of overhead confusing. Much time will be spent, and he will think he has accomplished nothing. However, when he is estimating a particular project, he will realize that his efforts have been rewarding.

Studies which have not been entirely completed may yield dividends. Once one has become conscious of the major items affecting overhead, he will automatically analyze the merits of jobs while estimating the cost of material and labor.

Many contractors have reaped their greatest benefits from overhead studies prepared by others. The contemplation of such work not only has yielded much valuable information but has served as a stimulant for further study.

The study of overhead is a *must* item in the electrical contracting business and cannot be put off.

CHAPTER 12

Can You Estimate Installation-only Contracts?

An *installation-only contract is defined as one in which the contractor supplies the labor, management, and installation services to install materials supplied by others.* In the trade, such contracts are often erroneously spoken of as "labor-only" contracts. The term labor-only not merely is erroneous but minimizes the value of the contractor's services. It must not be used when talking to buyers. The buyer must know that he is asking for more than just the labor of the mechanics.

Let us list some of the services that the contractor supplies along with labor:

1. Material service (for materials by others)
    a. Coordinating deliveries
    b. Checking materials—quantities and adaptability
    c. Storage facilities
    d. Protect materials—labor and material (tarpaulins, etc.)
2. Tools—consumed and depreciated
3. Cartage for tools and equipment
4. Testing and testing equipment
5. Field supervision
6. Construction engineering
7. Financing labor
8. Guarantee on work performed
9. Services of the contractor's general office force and equipment for timekeeping, accounting, and billing

10. Administrative management for coordinating all of the various functions

From the foregoing, we see that the contractor has many functions in connection with installation-only contracts. Among them is that of supplying a service in connection with materials supplied by others. In estimating, an item must be included to cover the cost of this service.

In addition to the material service cost, the estimator must include an allowance (excess labor) for protection against procurement failures and other installation-only hazards. By procurement failures is meant materials not ordered and delivered in accordance with the job requirements.

Electrical Estimating (McGraw-Hill Book Company, Inc.) lists the following installation-only project hazards:

1. Difficulty in getting information
    a. for installation plans
    b. on materials and equipment to be installed
2. Lack of authority to deal directly with vendors
    a. greatly delayed getting information
    b. resulted in getting wrong information
    c. made coordination of deliveries difficult
3. Materials ordered before being checked by contractor
    a. cables not cut and reeled to facilitate installation
    b. unbalanced quantities of material
    c. installation materials such as pipe supports, pipe fittings, mounting frames, etc., not in conformity with job requirements
    d. pipe fittings ordered for locations requiring pull boxes
    e. fittings and pipe supports not standard equipment
    f. pull boxes too small
    g. openings and knockouts in pull boxes and equipment wrong size or not in proper location
    h. fixtures (lighting) not adapted to the job
    i. fixtures (lighting) with wrong pipe entrances
    j. equipment normally assembled in the factory came to the job knocked down
    k. motor terminal heads inadequate
4. Materials not ordered because contractor had trouble getting complete listing
    a. installation materials—pipe hangers, mounting bolts, etc.

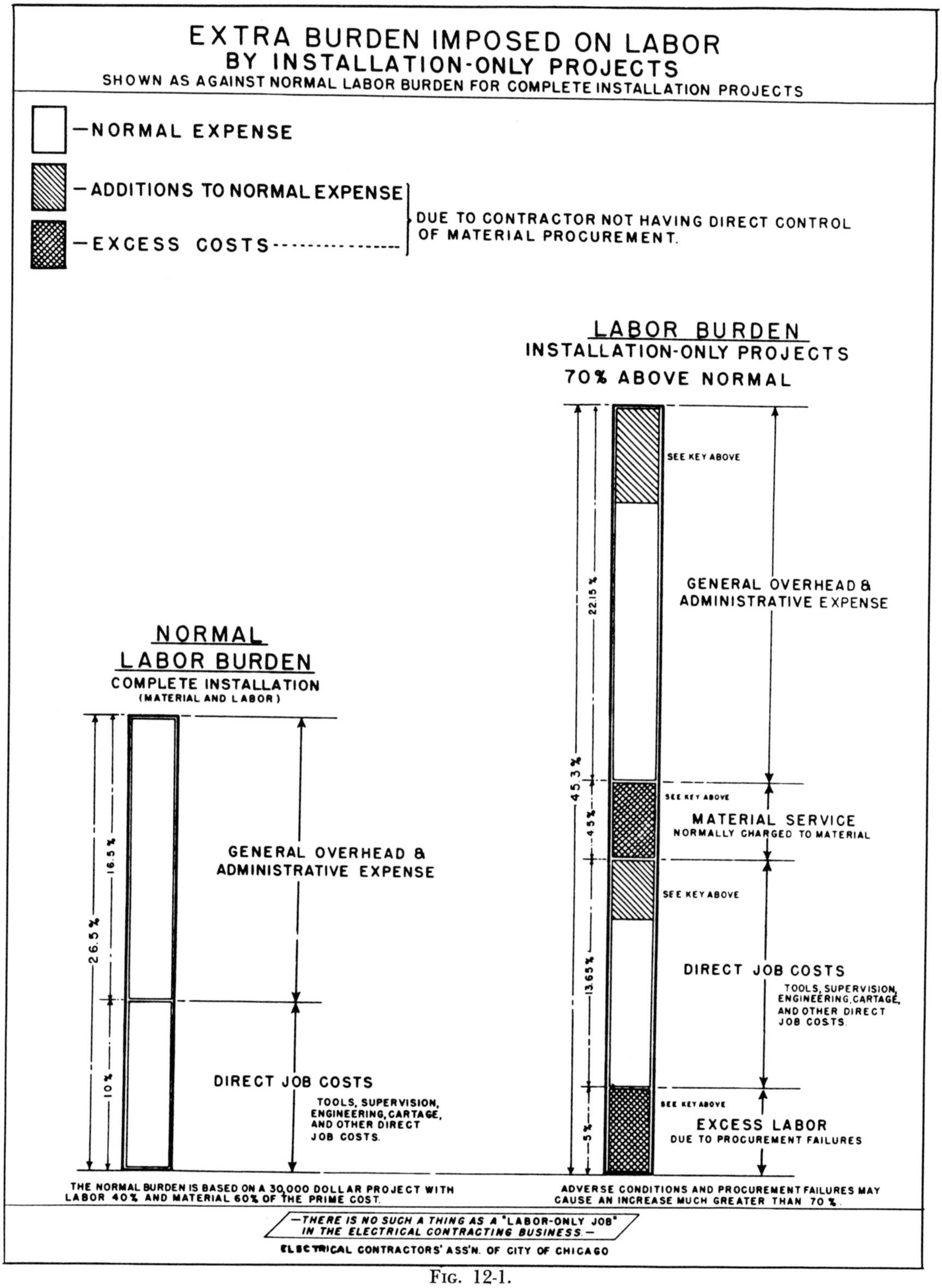
EXTRA BURDEN IMPOSED ON LABOR
BY INSTALLATION-ONLY PROJECTS
SHOWN AS AGAINST NORMAL LABOR BURDEN FOR COMPLETE INSTALLATION PROJECTS
—NORMAL EXPENSE
—ADDITIONS TO NORMAL EXPENSE
—EXCESS COSTS
DUE TO CONTRACTOR NOT HAVING DIRECT CONTROL OF MATERIAL PROCUREMENT.
NORMAL
LABOR BURDEN
COMPLETE INSTALLATION
(MATERIAL AND LABOR)
26.5%
16.5%
10%
GENERAL OVERHEAD & ADMINISTRATIVE EXPENSE
DIRECT JOB COSTS
TOOLS, SUPERVISION, ENGINEERING, CARTAGE, AND OTHER DIRECT JOB COSTS.
LABOR BURDEN
INSTALLATION-ONLY PROJECTS
70% ABOVE NORMAL
45.3%
22.15%
4.5%
13.65%
5%
SEE KEY ABOVE
GENERAL OVERHEAD & ADMINISTRATIVE EXPENSE
SEE KEY ABOVE
MATERIAL SERVICE
NORMALLY CHARGED TO MATERIAL
SEE KEY ABOVE
DIRECT JOB COSTS
TOOLS, SUPERVISION, ENGINEERING, CARTAGE, AND OTHER DIRECT JOB COSTS.
SEE KEY ABOVE
EXCESS LABOR
DUE TO PROCUREMENT FAILURES
THE NORMAL BURDEN IS BASED ON A 30,000 DOLLAR PROJECT WITH LABOR 40% AND MATERIAL 60% OF THE PRIME COST.
ADVERSE CONDITIONS AND PROCUREMENT FAILURES MAY CAUSE AN INCREASE MUCH GREATER THAN 70%.
—THERE IS NO SUCH A THING AS A "LABOR-ONLY JOB" IN THE ELECTRICAL CONTRACTING BUSINESS.—
ELECTRICAL CONTRACTORS' ASS'N. OF CITY OF CHICAGO

Fig. 12-1.

  b. pull boxes
  c. pipe fittings—insulating bushings, elbows, etc.
  d. mounting frames
  e. safety switches (list not complete)
  f. cable-splicing materials
  g. fixture aligners
5. Delivery of materials not properly timed
6. Jobbers substituted in place of materials ordered; materials not suited to the installation
7. Excess labor—protecting and shifting materials delivered long before needed
8. Excess labor for cleaning equipment that had been delivered far ahead of time
9. Excess labor spent on material and equipment brought in from other plants
10. Labor lost because of delays and disruptions resulting from the above procurement failures
11. Excess tool costs
  a. correcting equipment
  b. fabricating materials normally fabricated in the shop
  c. tools tied up on the job longer than they should have been
12. Owner's representatives did not cooperate as they should. Purchasing agents and other owner's representatives did not understand the contractor's problems
13. Plans supposed to be furnished by owner
  a. not available when needed
  b. not complete
  c. not construction plans
  d. not coordinated
  e. proper consideration not given to the building construction and layout of other equipment
  f. deceiving as to amount and nature of work

  It will be noted that there are 13 main headings and 35 subheadings. Such hazards may increase labor costs as much as 5 to 20% above normal. An allowance of 10% is generally recommended.

  Methods of estimating installation-only projects vary with the type of contract. Contracts are generally one of the three following types:

1. Fixed-price contract. The contractor agrees to a fixed price for supplying the labor and installation services for the complete project.

2. Cost-plus contract. The contractor bills for the actual cost of labor and installation services plus markups as stipulated in the contract.
3. Fixed-fee contract. The contractor bills for the actual cost of labor and installation services plus a fixed amount (lump sum) for markup.

Before preparing sample estimates, we shall review some figures regarding the relative costs of supplying labor and material. It was previously stated that, on a dollar basis, the cost of supplying labor is approximately 3½ times as great as the cost of supplying material. Some figures that verify the statement will now be studied.

Figure 10-3 provides a study of operating costs with separate figures for material and labor costs. Exhibit A lists direct job expenses, and Exhibit B lists the overhead.

We are principally interested in overhead ratios. However, in passing, the direct job figures can be studied. The totals of Exhibit A are 4.43% for material and 28.11% for labor. The 28.11% includes 14% for insurances. It has been previously pointed out that for accurate estimating, insurances should be treated as a material item when applying markups.

Subtracting the 14% (insurances) from the labor expense of 28.11%, we have 14.11%. The labor cost of 14.11% is approximately 3.2 times as great as the 4.43% for material cost.

The overhead percentages shown under Exhibit B, Fig. 10-3, are 6.96% for material and 23.89% for labor. Dividing 23.89 by 6.96 we get 3.43 plus.

Other studies provide reasons for selecting 1 to 3.5 as a representative ratio of material to labor overhead costs. Experiences of many have also justified this figure. However, the billing ratio is often changed because certain functions such as bookkeeping, billing, etc., are shifted to the field and billed as direct job expenses.

To illustrate the method of estimating, we shall assume that the work contemplated would normally have a 60/40 M/L ratio with estimated costs of $6,000 (60%) for material and $4,000 (40%) for labor.

Turning to Fig. 11-2, we find that the normal labor overhead for a $10,000 installation is given as 26.2%. In our estimate 26% will be used.

First we shall study the estimate for a fixed-price contract and base our figures on normal experiences. Later, we shall prepare a second estimate for a project with minimum hazards.

Estimate 1. Normal Experience

	Job Expenses	Labor
Estimated labor (normal)...................		$4,000
Excess labor (10%).......................		400
		$4,400
Tools—3% of $4,400......................	$ 132	
Miscellaneous job expenses—3% of $4,400....	132	
Insurances—12% of $4,400.................	528	
Material service 4.5% of $4,000.............	180	
Totals..............................	$ 972	$4,400
Overhead......................... (10%)	97 (26%)	1,144
	$1,069	$5,544
Return........................... (5%)	53 (10%)	554
Estimated sell incidentals..................	$1,122	$6,098
Estimated sell labor.....................	6,098	
Estimated sell the job...................	$7,220	

Excess Labor

In the foregoing estimate there is an item of 10% added for excess labor. Causes of excess labor are numerous. The hazards of installation-only projects, previously listed, are all possible causes of increased labor costs.

In a previous chapter, effects of untimely deliveries of materials were studied. In addition to the untimely delivery of materials, there is also the possibility of materials not being adapted to the work. The experienced estimator is acquainted with the costs and delays resulting from getting the wrong materials. For those not so familiar with electrical construction, two of the items listed as hazards will be discussed, namely: (1) cable not cut and reeled to facilitate installation and (2) motor terminals inadequate.

As an example of cable not being properly shipped to facilitate installation, we shall take a case where the cable for two three-wire feeders is delivered on a single reel. Three lengths for each feeder must be run off, cut, and put on separate reels. In addition to the labor involved, there may be delays while waiting for additional reels to be delivered.

It is quite common to find terminal heads on large motors inadequate. The mechanic does not learn of this until the motor has been delivered. There is not only the labor and expense of replacing the terminal head, but there is the delay and interruption of the work.

The potential hazards of each installation-only project must be studied carefully. The estimated excess labor may be more or it may be less than 10% depending on the particular job.

There are many things to be considered when the estimator is establishing excess labor charges. The experience and the attitude of the men procuring the material and the attitude of the vendor supplying it are important factors. The type of installation is also a factor.

A firm may have several typical installations at different times. Chances are that the excess labor will be much greater on the first installation than it will on those that follow.

Direct Job Expense

The direct job expenses and material service costs have a separate listing, although they are figured as a percentage of labor. This is done because they do not require the same markups as labor. For markup purposes, they are treated as items of material.

If a contractor was preparing a preliminary estimate, or if he felt that conditions warranted it, he would more than likely list job expenses with labor and use a common markup. In the foregoing example, if the job costs had been listed with labor and subjected to the same markups, the final estimated cost would have been $226 greater, or $7,446 instead of $7,220.

Material Service

In the early part of this chapter, material services supplied by the contractor were listed. The expense of supplying such services is usually estimated to be approximately 3% of the cost of material. Had 3% of the material cost been used in our estimate, we would have had 3% of $6,000, or $180. In the estimate, 4.5% of $4,000 (the cost of labor) was used and the result was also $180. Naturally, on a project with an M/L ratio of 60/40, 4.5% of the labor cost would be the same as 3% of the material cost.

For an installation-only project, one may not care to take the time to estimate the material costs. If a percentage of labor cost for material service is used, the estimate can proceed without the exact cost of material being known. The result may not be 100% correct, but at best, estimates of material service costs are only approximate.

Material service costs, like excess labor, vary with the type of project and deliveries. Again, looking at the list of hazards, we can see many things that could effect the over-all costs of handling materials.

Return

The term *return* is used instead of *profit* for two reasons:

1. It reminds the contractor that the amount being added will not be realized as a profit unless all costs have already been covered.
2. It is often useful for selling purposes.

Selling

Under the heading "Propositions for electrical work," we shall learn how valuable an estimate may be as a selling aid. Some contractors still think that buyers should not see the estimates for their work. However, most contractors are beginning to realize that, if their estimate is right and they can explain their charges, their chances of selling are better if the buyer can see the figures. This does not mean that the estimate should be turned over to the buyer unless special conditions exist. Normally, the estimate is strictly the property of the contractor and must remain in his hands.

If the foregoing estimate was to be used for selling, it would be well to have only one listing for labor. It would be "Estimated labor—$4,400." The buyer would be informed verbally that the labor cost could be reduced if the contractor supplying the installation labor could have charge of the material purchases.

Briefly, the things that the buyer could learn from our estimate are:

1. More than just the base cost of labor is involved.
2. Low markups are applied to incidental costs.
3. The amount added for "return" is not excessive.

Buyers have asked contractors why they used the term *return*. Experiences show that the reaction of buyers is very good when they are told that the amount added for return is an anticipated profit but, owing to the hazards involved, may not be realized as such. It seems to be natural for one to be happy when he learns that a supplier is not going to make too much on his business.

Let us set up another estimate for the same work and assume that conditions are especially favorable, the reason being that it is a duplicate project and the procurement of materials will be in the hands of capable and experienced men. It is expected that all materials will be adapted to the job and deliveries timely.

Estimate 2. Conditions Exceptionally Favorable

	Job Expenses	Labor
Estimated labor (normal)...................		$4,000
Excess labor (3%)........................		120
		$4,120
Tools—3% of $4,120......................	$ 124	
Miscellaneous job expenses—3% of $4,120....	124	
Insurances—12% of $4,120.................	494	
Material service—2% of $4,000.............	80	
Totals...............................	$ 822	$4,120
Overhead.......................... (10%)	82 (25%)	1,030
	$ 904	$5,150
Return............................ (5%)	45 (10%)	515
Estimated sell incidentals..................	$ 949	$5,665
Estimated sell labor......................	5,665	
Estimated sell the job...................	$6,614	

The estimated sell price for Estimate 1 was $7,220 and for Estimate 2, $6,614. The second estimate is $606 less than the first. The difference is almost 8.5%, which, in competitive bidding, would represent a comfortable margin.

It is generally agreed that installation-only work is not the most desirable, but if one is going to bid on it, he cannot afford to overlook any advantages that a particular project offers.

COST-PLUS CONTRACTS

We are in the habit of thinking that a cost-plus contract includes the material along with the labor for installing it. However, there are I-O (installation-only) contracts let on the cost-plus basis. Requests from buyers of I-O work usually come in one of the following forms:

1. Total cost per hour for labor including insurances, tools, cartage, and all incidentals required for labor plus markups for overhead and profit
2. Bid for labor only to install materials and equipment supplied by others

It is the more experienced buyers that use the first form. We shall prepare an estimate for such a request. The new estimate will be marked No. 3, and numbers will be assigned to the various items for use in later references.

For our estimate, we shall assume that the complete cost (labor and material) of the project is $30,000, with $18,000 for material and $12,000 for labor. Again we shall prepare the estimate first and let the discussion follow.

1. Estimate 3. Installation Labor and Services for a $30,000 ($18,000 Material, $12,000 Labor) Project

2. Base cost of labor—journeyman rate	$3.00 *
3. Prorata foreman's extra pay	0.06
	$3.06
4. Overhead (25%)	0.765
	$3.825
5. Insurances (12% of $3.06)	0.367
6. Tools (3% of $3)	0.09
7. Direct job expenses—miscellaneous (3% of $3)	0.09
	$4.372
8. Return (10%)	0.437
	$4.809 †

* Journeyman rates in many cities now exceed $4.
† If the contract was inviting, the contractor would more likely bid $4.80 per hour.

1.* The estimator must know the approximate size of the project in order to select the proper overhead percentage for markup.

2. The journeyman electrician pay is $3 per hour, and it is used as the base rate.

3. In the estimate, 6 cents has been added to the journeyman rate to compensate for the extra pay received by the foreman. This is necessary because the foreman's time is to be billed at the same rate as the journeyman (see "Apprentices" later).

It is assumed that there will be one foreman for every eight journeymen and that the total addition is 9 (eight journeymen and one foreman) times 6 cents, making a total of 54 cents. This allows a rate of $3.54 per hour for the foreman's pay. If the foreman's rate was $3.36 per hour, the addition would be 4 instead of 6 cents.

Apprentices

If apprentices were to be used along with the journeymen, more than likely the contractor would use the $3 rate without adjustment. The premium paid the foremen would be offset by the saving on apprentices.

* Figures refer to parts of Estimate 3.

Separate Prices

If agreeable to the buyer, the contractor may quote separate prices for foreman, journeyman, and apprentices. There would then be a separate estimate for each rating. The method of estimating would be as shown by Estimate 3, except that each base rate would be used without adjustment. With a single estimate, the estimator could establish a flat percentage to be added to each base rate. In our estimate the total addition to the average pay of $3.06 was 57%.

Excess Labor and Material Service

The excess labor is billed for by the hour along with the normal labor. Most of the material service is taken care of by the mechanics and is also billed for by the hour. Excess office expenses are compensated for by the overhead markup on excess labor and material service labor.

4.* The overhead used in Estimate 3 is less than that used for Estimates 1 and 2 because the project is three times as large.

5. The insurances and other job costs are added after the labor overhead has been applied. Only 10% markup is applied to these items, and it is marked "return." Strictly speaking the 10% ($0.437) is not all return (anticipated profit). The contractor knows this but is not splitting hairs when he knows there are many things that will influence his costs. The overhead markup on labor is sufficient to cover the overhead on incidental expenses if the work progresses smoothly.

If, for any reason, one feels that he should be more exacting, he can make a separate listing of job expenses and use 5% plus 5% for markups. On a small job the markups would be 10% plus 10%.

6. Tool costs have been discussed at length in previous chapters.

7. In the discussion of the relative costs of supplying material and labor, we had a figure of 14.11% for direct job expense on labor. In Estimate 3, 3% was used. A study of Exhibit A, Fig. 10-3, will reveal that there are some items listed that can be absorbed by labor on the job and others will be the responsibility of the owner. Supervision and timekeeping can be absorbed by the job labor. Drafting, blueprinting, and inspection must be paid for by the owner. Each job has its individual requirements.

8. We have already discussed the item of "return." As stated before, on a large project, the overhead markup on labor may be suf-

---

* Figures refer to parts of Estimate 3.

ficient to cover the job expenses as well, and the 10% markup can all be enjoyed as return.

We now come to the second request for a bid on a cost-plus I-O project. The buyer may ask for an hourly rate without stipulating what that rate is to include, or he may just ask for a bid on labor. In either case he has little idea of what is needed besides the actual work of mechanics.

The contractor knows his costs, and he knows how much he must charge for labor and installation services. His big problem is that of devising a method of presenting his proposed charges so that they will be attractive to the buyer. This gets into the writing of propositions, which is a subject covered in a later chapter.

FIXED-FEE CONTRACT

Methods of establishing a price for a fixed-fee project vary according to the practices of the individual contractor, size of project, type of project, and type of request for bids.

Size of Project

Up to a certain point, the larger the project, the less (in percentage) the cost of handling. Generally, fixed-fee bids are requested only for very large projects.

Type of Project

An industrial plant of more or less standard type costs less than a generating plant or other installations requiring a great deal of special equipment. The former costs less to figure, moves more rapidly, and requires less attention throughout the life of the job.

Methods of Management and Billing

These two items have been listed together because they both belong to the overhead group and can be illustrated in the same example. To illustrate how these functions can affect the fixed-fee price, two types of requests for bids can be studied: (1) The contractor is to bill labor at payroll cost and to submit a fixed fee to cover all other costs of labor and installation services; (2) the contractor is to bill all costs directly and submit a fixed fee for management services.

Estimates for a $500,000 (base cost), 60/40 I-O project can be made. The estimated cost of material would be $300,000 (60%), and the estimated cost of labor $200,000 (40%). The normal is $200,000. Adding 10% for excess labor we have $220,000.

Can You Estimate Installation-only Contracts? 115

Estimate 4a. Fixed Fee with All Job Expenses Included

	Job Expenses		Labor Expenses
Tools: 3% of $220,000	$ 6,600		
Miscellaneous job expenses: 3% of $220,000	6,600		
Insurances: 12% of $220,000	26,400		
Material service: 3% of $200,000	6,000		
Total insurance and incidental expenses	$ 45,600		
Overhead (5%)	2,280	14% of $220,000	$30,800
	$ 47,880		$30,800
Return (5%)	2,394	10% of $250,800 *	25,080
Sell price incidentals	$ 50,274		$55,880
Sell price labor expenses	55,880		
Sell price the job	$106,154		
Fixed-fee bid $106,150			

* The $250,800 includes $200,000 normal labor, $20,000 excess labor, and $30,800 overhead

On the type of contract illustrated by Estimate 4a, the contractor must watch his labor closely. Although the payroll is billed to the buyer, the contractor has to stand the expense of insurances and other job costs. In our next estimate we shall deal with the type of contract in which the contractor is not so vitally concerned in the total hours of labor.

We now come to the second type of fixed-fee contract where the buyer is billed directly for all expenses except a prorata share of executive salaries and the small prorata expense of the home office. It is strictly a cost-plus contract with a fixed-fee. The project manager, engineer, bookkeeper, and all other required personnel are on the job, and their salaries billed directly. Tools, cartage, insurances, and all other expenses are also charged directly to the job cost.

Normally, it would not be good management to let a $500,000 contract in the manner just described. However, since we are principally interested in the method, the base figures used in the first estimate will be employed.

We must establish a charge for management costs. Exhibit *B*, Fig. 10-3, lists administrative salaries as 8.35% of labor cost. This is for entire time given over to many jobs aggregating $500,000. In our problem, there is one project amounting to $500,000 and there is a project manager. Two per cent should be ample.

The home office has practically no expense with the project, but some allowance must be made. With an estimated total cost of $220,000 for labor, 2% would be $4,400. This should be more than ample if other business could go on as usual. However, we must make some allowance for possible interference.

Using 2% for executive salaries and 2% for home-office expense, we have a total estimated overhead of 4%.

We have been using 10% as a fair return on labor. It is not too great for any type of project, but here we shall try to stay close to what we think contractors would use in competition and use 5%. With all expenses charged directly to the project, the 5% for return could safely be called profit.

Manning the job office must be contemplated from the standpoint both of costs and of securing the proper personnel. If the contractor cannot spare people from his regular organization, he must be sure that he can get them elsewhere when and if needed.

As a part of the total cost of the project, the contractor must include the estimated expense of manning, equipping, heating, and lighting the job office. This expense is largely dependent on the duration of the contract.

We are using a base rate of $3 per hour. Allowing for the extra pay of foremen, we shall say that the average pay is $3.06 per hour. At this rate, the total labor cost of $220,000 would represent approximately 7,200 hr. Turning to Fig. 8-1 we see that the economical time for a 7,200-hr project is between 53 and 57 weeks. From this the contractor knows that he must figure that salaries and expenses will run a year or more.

To avoid prolonging the discussion, we shall assume that the contractor has estimated the total cost of the job office, including salaries, equipment, supplies, and other expenses, to be approximately $35,000. This is a little over 15% of the total payroll. We shall use 15%.

With the use of the percentages just established along with certain standard cost percentages, the estimate is as follows:

Estimate 4b

Total payroll (including excess labor)	$220,000
Job office—salaries, equipment, etc. (15%)	33,000
Tools (3%)	6,600
Miscellaneous job expense (3%)	6,600
Insurances (12%)	26,400
Estimated total billing	$292,600
Management and general office expenses (4%)	11,704
Total estimated cost	$304,304

Estimate of Fixed Fee

Management and general office expenses	$ 11,704
Return—5% of $304,304	15,215
Total	$ 26,919

Fixed-fee bid $27,000

The estimate serves only as a guide. How much the contractor finally decides to bid depends on how well the project will fit in his program, the type of project, the general contractor, and many other things.

We could go on indefinitely setting up methods of establishing fixed-fee bids, but the foregoing estimate includes the two essential items —management costs and return. As stated before, every item of expense, except management from the main office, is billed directly to the owner. Therefore, as shown by the estimate, there are only two items to be included in the fixed-fee bid—cost of management and return.

CHAPTER 13

Can You Write Proposals?

Do you know that the type of proposal may be responsible for obtaining or losing the contract? This is particularly true when bidding for cost-plus contracts or for fixed-price contracts that are not completely covered by plans and specifications.

BIDDING ACCORDING TO PLANS AND SPECIFICATIONS

The simplest form of bidding is that for work completely covered by plans and specifications. There is little required besides the bid price and the usual statement about the work being installed in accordance with plans and specifications submitted.

In some cases it is well to include the dates appearing on the plans and specifications. This is especially true when there have been revisions.

In any event, the contractor must make sure that he keeps a record of all dates and that all plans and specifications are clearly marked for future identification.

If a contractor is bidding to an out-of-town firm and feels that information regarding his organization should be supplied, such information must be on a separate sheet attached to the proposal.

It is often well to supply the estimated time required for completion. Completion dates will be discussed later.

THE PROPOSITION

Proposition Type 1

The standard form of proposal, when bidding according to plans and specifications, is as follows:

> *For the sum of ONE HUNDRED AND SEVENTY-FIVE THOUSAND DOLLARS ($175,000),* we propose to furnish material and labor for the installation of ELECTRICAL WORK in the Kellar Office Building to be erected at 1220 North Oak Street.*
> *All work to be installed in accordance with plans and specifications submitted by you.*

If advisable to give the estimated completion date, the following can be included:

> *With normal working conditions, the time required for completion will be twenty-five weeks.*

The contractor includes the qualifying statement *"With normal working conditions."*

Reasonably clear working spaces, usual progress of the job, an 8-hr working day, and 5-day weeks are factors belonging to normal working conditions. If the job does not progress as it should and normal conditions do not prevail, the contractor can write a letter to the owner calling attention to such conditions thereby relieving himself of the obligation to complete the work in the time set forth in his proposition.

A Dangerous Practice. A dangerous practice among some contractors, when asked to give a completion date, is to say "As required."

The contractor using the statement "as required" is putting himself in a bad position. Other trades may work overtime, and he will also have to work extra hours to keep from having his work covered up. Chances of being able to collect for the overtime hours under such conditions are very doubtful.

It is a regular thing for owners to try to get contractors to overman their jobs. The "as-required" statement would give them extra leverage.

---

* There are cases in which a contractor may choose to omit writing out the proposed price.

BAIRD AND KAIN—ELECTRICAL CONSTRUCTION

2424 Clifton Street, Chicago, Illinois
Tel. Mead 2-2424

May 21, 1960

Plunket and Plunket
Architects
2810 Michigan Road
Chicago, Illinois

Dear Sirs:

For the sum of ONE HUNDRED AND SEVENTY-FIVE THOUSAND DOLLARS ($175,000) we propose to furnish material and labor for the installation of ELECTRICAL WORK in the Keller Office Building to be erected at 1220 North Oak Street, Beman Hills, Ill.

All work to be installed in accordance with plans and specifications prepared by you and dated April 1, 1960.

Respectfully submitted,

H. R. Baird, president

FIG. 13-1.

Proposition Type 2

A proposition requiring dates would be similar to the following:

For the sum of NINETY THOUSAND DOLLARS ($90,000) we propose to supply the electrical installation in the addition to the Beverly Hospital located at 1020 Pine Street.

Work to be installed in accordance with the following plans and specifications supplied by you:

Plans E-1 to E-7 incl., dated Aug. 1, 1958 with revisions. Latest rev. date Oct. 2, 1958.*

Specifications pages 1 to 42, dated Aug. 1, 1958 and Addendum No. 1, dated Oct. 2, 1958.*

With normal working conditions and normal progress of the job, work can be completed in 20 weeks.

There are contractors who garnish their proposals by starting out with "We are pleased to have an opportunity to quote you . . ." and ending with "We will be happy if favored with your contract." That

* If the plans and specifications are numbered, such numbers must be included.

is so much bunk and wasted energy. The buyer does not give a hoot how pleased the contractor is. He is concerned in the price, and about all he will look at in the proposal will be the figures following the dollar sign.

COMPLETION TIME

Before taking up the next form of proposition it will be well to complete the discussion on "completion dates."

Many contractors try to avoid giving completion dates, whereas others make a practice of regularly supplying estimated time for completing contracts.

Studies have brought to light many cases where contractors had benefited by supplying completion dates with their proposals. No case was found where a contractor had suffered from such a practice.

Completion dates have protected contractors from increase in labor rates, unusual price rises, and extended duration of the job. For costs of "extended duration," see *Electrical Estimating* (McGraw-Hill Book Company, Inc.).

To be able to estimate completion dates, one must have a good knowledge of building construction, ample tools and equipment and know where he is going to get suitable mechanics to do the work. The labor market is always a hazard to be considered. Contractors seldom overlook this.

Figure 8-1 provides a table of optimum durations for industrial projects within a certain range. This table has been used as a guide for estimating completion dates for many types of projects.

Buyers are always glad to be given completion dates. They have large investments in their plants and want to start operations as soon as possible. In most plants, little can be accomplished until the electrical installation is ready to operate the machinery.

FIXED-PRICE CONTRACTS—NO PLANS

The most unsatisfactory type of bidding is that for work with neither plans nor specifications and competition to be considered. One never knows whether or not he is getting the same verbal instructions as his competitors, and he is not sure that his competitors are going to figure the same type of high-class installation.

Propositions for such work are usually lengthy and require much care and time in preparation. They amount to a combined specification and bid. And when owners receive them, they look at little besides the price. Nevertheless, the contractor must include details to avoid later disputes.

No real solution has ever been found to right this condition. Contractors have ideas of what ought to be done but never do it. Besides, each case represents an individual problem.

As a suggestion: one might do well to omit details and just submit an itemized list of costs as follows:

New main distribution panel	$1,040
Processing-room panel	248
Pump-room panel	220
Feeder to processing room	210
Feeder to pump room	320
Power branch wiring	520
Motor starters and motor connectors	333
Insurances, cartage, and miscellaneous	98
Total cost of work	$2,989

The following note to be added:

Work will not only comply with all local and national code requirements but will be installed in accordance with the best engineering practices.

The design of panels and capacity of feeders provides for 20% additional load.

If offered the contract, the bidder can submit a detailed description of the work for approval. This to be given to the buyer to look over before the contract is signed.

The buyer would read the listing because he is interested in individual costs. He would more than likely read the notes to see what qualifications followed.

The listing and the notes would help the bidder gain the confidence of the buyer. They might also make it easier for him to get a chance to sell the job firsthand.

The subject of contractors being opposed to revealing any of their detailed costs is treated in other chapters.

COST-PLUS BIDDING

For cost-plus contracts, the better engineering firms provide details for bidding. This helps ensure uniform prices.

Few individuals have any idea of how to go about getting cost-plus bids. They just ask contractors to give them prices. As a result, an assortment of prices such as the following may be received:

1. Labor and material at cost plus 20% and 10%
2. Cost plus 15% and 10%
3. Material at cost plus 10% and 10% and labor at cost plus 20% and 10%
4. Material at cost plus 10% and labor at cost plus 30% and 10%

Such bidding is confusing to buyers. Among uninformed owners, the better bids are often overlooked.

The reader should know that contractors bidding to regular customers often vary their bidding to suit the wishes of the individual client. Here, we are studying bids for new accounts.

When asked to bid for a cost-plus contract where no bid form is provided, a contractor has much to gain by calling the buyer and asking him what type of bid he would like to have. This will start the buyer thinking, and one may be able to get an interview. An interview enables one to get acquainted and is often the means of selling the contract.

The following is a recommended form for bidding cost-plus work: *

BAIRD AND KAIN—ELECTRICAL CONSTRUCTION
2424 Clifton Street, Chicago, Illinois
Tel. Mead 2-2424

June 1, 1960

Black and Black
Hardware Specialties
4920 Edge Street
Chicago, Ill.

Dear Sirs:

We submit the following prices for installing the electrical work in your plant at 5020 Belmont Ave.:

Material will be billed at cost plus 10% for material service and overhead, and 5% for service return.

Labor will be billed at cost plus 30% for service charge and overhead, and 10% for service return.

Insurances, inspection, cartage, and incidental job expenses will be billed at cost plus 10% for service charge.

Work to be installed as directed by you. You will be supplied with high-grade construction engineering service, top-rate mechanics, and careful supervision for your installation.

Respectfully submitted,

H. R. Baird, president

* Percentages of markup to be varied according to the wishes of the individual contractor.

The foregoing is submitted as a form only. The contractor must study all conditions and vary both form of bid and bid price according to the merits of the individual job. In the chapter on overhead we learn that costs vary according to the type and size of installation and the length of duration.

In general, contractors do not quote an hourly rate, including insurances, job expenses, and markups, for labor. Buyers find the total amount startling because they are in the habit of thinking of bare labor costs without incidentals and markups.

If the contractor is asked to submit an hourly rate, it is strongly recommended that he substantiate his price by supplying a listing of costs involved. The accompanying listing, "Analysis of Labor Costs," illustrates a method recommended by the author several years ago. It is now regularly used by many contractors across the country.

One must always have an itemized listing of costs available. Such lists are needed to substantiate billing as well as quoted prices. Buyers are aware of many of the costs involved, but few realize how such costs mount when added up.

A sizable project may require as much as 50% (including all costs and profit) addition. In many localities, the present rate of journeymen electricians is close to $4 per hour. Add 50% and you have $6. That is 10 cents per minute. Twenty minutes spent getting out tools and going to the place of work represents $2 cost before a stroke of work is done. Is it any wonder that buyers who started out when rates were $1.50 per hour find such costs startling?

In a chapter on selling, we shall learn of other useful information to be supplied when billing or bidding for work. As stated before, propositions vary greatly depending on the contractor, buyer, job conditions, and many other factors. Now we shall study one of the special cases.

A SPECIAL CASE

Figures 13-3 and 13-4 provide a proposition and an estimate which are close facsimiles of those sometimes used by a well-known and prosperous electrical contractor. This is only one of the many examples of his work that could be presented. His bidding is adapted to various conditions. The particular examples are chosen because they promote much discussion.

Figure 13-3 has letters and Fig. 13-4 has numbers in the margin. These are to serve as a guide for one reading the text.

ANALYSIS OF LABOR COSTS
Average cost of items to be added to a $3.75 wage rate (hourly), based on an annual volume of $550,000 (cost) with a "60/40" material-labor ratio

Direct Job Costs   Per hour

1.0% Association dues	$0.0375
2.7% State unemployment contribution	.101
0.3% Federal unemployment contribution	.0113
2.5% Workman's compensation insurance	.0937
1.5% Federal old-age security	.0563
1.5% Public liability, property damage, occupational disease, etc.	.0563
5.0% I.B.E.W. benefit fund (El. Ins. trustees)	.1875
3.0% Tools and construction equipment (consumed and depreciated)	.1125
3.0% Engineering (varies according to type of project)	.1125
20.5%	**$0.7686**

Overhead Expense

2.0% Rents—office and storeroom (including light and heat)	$0.075
2.0% Furniture, telephone, stationery, and postage	.075
1.0% Taxes, licenses, bonds, legal and miscellaneous expenses	.0375
7.0% Salaries for office and storeroom employees	.2625
2.0% Travel and sales expense (automobile, railroad, etc.)	.075
8.0% Administrative salaries	.300
1.0% Adjustment factor (3 bad years in 10)	.0375
23.0%	**$0.8625**

Base pay—Class A journeymen electricians		$3.75
Direct job costs	20.5%	0.77
Overhead expense	23.0%	0.86
Totals	43.5%	$5.38

Notes:

1. Percentages are for illustration purposes only. They must be checked locally.
2. For installation-only projects (materials by others), a material service charge must be made.
3. To determine the hourly charge for apprentices and foremen, add 43.5% to the base rate paid same.
4. No allowance has been made in the above listing for service return (anticipated profit).

Fig. 13-2.

The Proposition (Fig. 13-3)

*A.** The contractor quotes a price based on an "approximate estimate," a copy of which accompanies the proposition. Buyers are interested in detail figures.

A PROPOSITION FOR ELECTRICAL WORK
(A special case. Names and addresses omitted.)

Dear Sirs:

A   We wish to submit an approximate estimate for an upset price of ELEVEN THOUSAND FOUR HUNDRED AND FIFTY-FOUR DOLLARS ($11,454) for the electrical work to be installed in department 7, building 3, of your plant at the above address. All work to be installed and billed for as follows:

B   Billing for this work to be according to our standard basis for this type of installation.

C   Material will be billed at cost plus 10% for material services and return. Duplicate copies of jobbers' invoices for material will be attached to each bill.

D   Labor will be billed from timecards approved and signed on the job by your representative; 20% will be added to labor cost (payroll) for construction engineering, general overhead, and administrative expenses, and 10% will be added for service return.

E   The estimate includes the installation of the following:

    Two 15-hp, three 10-hp, and five 5-hp motors
    Six 200-watt RLM fixtures and 30 80-watt fluorescent fixtures
    10 SP switches and 20 duplex receptacles

F   Certificate of electrical inspection to be obtained and paid for by us. In addition to complying with all rules of the local and national codes, work will be installed in accordance with the best engineering practices.

G   Feeders, branch circuits, and panels will be in excess of code requirements to take care of minor additions.

The cost of this work is guaranteed not to exceed $11,454.

Respectfully submitted

FIG. 13-3.

The estimate, although conservative, is high enough to allow some leeway. If minor changes are made, no extra charges will be necessary. Perchance all goes well and there are no extra charges; the work can be completed for less than the quoted price. This will please the buyer, and it will be easier for the contractor to sell the next job.

* Italic letters refer to parts of Fig. 13-3.

B. The statement "Billing for this work to be in accordance with our standard basis . . . " ensures the buyer that he is to be billed at rates in keeping with those used for other customers. It may also leave the impression that rates for billing electrical work are more or less standard among contractors.

The qualifying statement "for this type of installation" leaves the way open for other types of billing when required.

C. The contractor uses one percentage (10%) for combined service (overhead and management) charge and service return (anticipated profit).

Buyers do not like to have contractors realize large profits on the material furnished for their work. With a combined markup of 10% for both overhead and profit, it is obvious that there cannot be much included for either.

The low markup informs the owner that he has little to gain by buying materials from suppliers and having the contractor install them. Besides he is told that he will be given duplicate invoices for material delivered to the job.

D. The timecards are to be approved by the owner's representative. The owner may have complete confidence in the contractor, yet it gives him a feeling of satisfaction to know that there will be a check against any possible error in bookkeeping or billing.

E. A brief outline of the principal items to be installed provides a record and may cause the buyer to check more closely to see that none of his requirements have been overlooked.

F. The contractor calls attention to the fact that compliance with codes does not ensure that work will be installed in accordance with the best engineering practices.

G. The buyer is informed that the price quoted is for an installation that more than meets the bare requirements of his plant. It provides some capacity for future additions.

In competition the extra allowance for future additions may help sell the contract. However, we must admit that in many cases, liberal allowances have been responsible for losing contracts.

In bidding to regular customers, the contractor can consider the possibilities of future additions because he will have a chance to explain his estimate. In bidding against competition, he has to be careful that he does not figure himself out of a job. It is better to wait and recommend additions after the contract is signed.

The reader may question the advisibility of submitting duplicate invoices for material. This has long been debated among contractors.

Many think it is better to have the customer know nothing about the prices of material.

In these times, large plants have little difficulty securing material prices. They usually have some supply-house representative knocking on their doors and wanting to sell directly.

The contractor who supplies duplicate invoices saves himself a lot of billing and often gains greater confidence of his customer.

The Estimate (Fig. 13-4)

1.* The idea of supplying buyers with detailed estimates of cost was a "horror subject" among contractors for many years. They thought that supplying such information would be undermining their business. The courageous have opened up the way for new thinking.

Estimates are not commonly provided with proposals except (*a*) when bidding to regular customers, (*b*) at the request of the buyer, (*c*) when the contractor's better judgment tells him it is prudent.

2. It will be noted that the price of conduit includes fittings. The contractor does not reveal any exact prices, yet he supplies the owner with the information wanted, i.e., the individual costs that are responsible for such a large total. Prices are substantial enough to provide some leeway.

3. In an estimate of this type, an item must be included for incidentals. The amount is varied according to the nature and size of the project.

4. The estimate as well as the proposition reveals the low markup used for materials.

5. As a rule, contractors quote prices for journeymen and foremen only. The saving on apprentices is consumed by the above-average journeymen. There are times when buyers question such billing and want to know why the journeyman rate is charged for apprentices.

To forestall any question, the estimate includes the classification and rate of all mechanics that are to be used for the work. The listing also forestalls any dispute regarding billing for the superintendent's time. Some buyers maintain that the superintendent's time belongs to the overhead expense.

6. Here the cost of tools and insurances is lumped together. This is unusual, as it is common practice to estimate and bill each as a separate item.

Note that no overhead is applied to the cost of insurances and tools.

* Numbers refer to parts of Fig. 13-4.

PRICING SHEET   DATE OCT. 1, 19__

CHI. ELECT. EST. ASSOC. FORM 4

JOB OR BLDG. HARTMAN MILLS   LOCATION
BID TO DEPT. 7, BLDG. 3   ADDRESS
ARCHT. OR ENG'R.   ADDRESS
PLANS MARKED   SCALE   SPEC. No.   EST. No.
EST. BY   PRICED BY   EXTENSIONS BY   CHECKED BY   S.O.No.   SHEET No.

	MATERIAL	QUANTITY	MATERIAL UNIT	EXTENSION	LABOR UNIT	EXTENSION
(1)	~ PRELIMINARY ESTIMATE ~					
(2)	2" GALV. CONDT. - INCL. FTGS.	120		90 —		
	3" " " " "	60		93 —		
	No. 4 "RHRW" WIRE	2000'		318 —		
	No. 3/0 " "	850'		460 —		
	#500,000 CM. "	240'		384 —		
	20 CT. LTG. PANELS	2		210 —		
	200 W. "RLM" FIXT. COMPL.	6		52 —		
	80 W., 2LP. FLUOR. FIXT. "	30		920 —		
(3)	MISCL. MAT. & CONTING.			130 —		
	INSPECTION FEES			110 —		
	CARTAGE & CONSUMED TOOLS			48 —		
	TOTAL MATERIAL			3840 — *		
(4)	MATERIAL SERVICE & RETURN	10%		384 —		
	MATERIAL SELL PRICE					4224 —
	LABOR - APPRENTICE	260 HRS @	2 50	650 —		
(5)	CLASS "A" ELECT.	780 " @	3 80	2964 —		
	FOREMAN	260 " @	4 00	1040 —		
	SUPERINTENDENT	50 " @	4 30	215 —		
	TOTAL LABOR (COST)			4869 —		
(6)	INSURANCES & DEPR. TOOLS	15%		730 35		
(7)	CONSTRUCTION ENG. & OVERHEAD	20%		973 80		
				6573 15		
	SERVICE RETURN	10%		657 30		
						7230 45
	TOTAL SELL PRICE					11454 45
(8)	GUARANTEED MAXIMUM COST			11454 —		
(9)	BILLING TO BE FROM ACTUAL COST RECORDS.					
	* INCLUDES COSTS NOT SHOWN.					

FIG. 13-4.

7. Again we have the cost of two items—construction engineering and overhead—combined. It is all a part of the program to avoid having any part of the estimate seem excessive. The presence of the "construction engineering" item detracts from the 20% which is principally overhead.

A reader familiar with operating costs for electrical work will note that all the markups used are very modest.

8. Both 8 and 9 are a repetition of statements made in the proposition. It is well to impress these two facts on the mind of the owner.

GENERAL

The foregoing special proposition was not given as any standard to be followed. Many of the features would not be adapted to everyday practice. It was chosen because it provided many items of interest.

The subjects of propositions and billing have been neglected by many contractors and studied carefully by many others. Those who have studied the subjects have been rewarded often for their efforts.

It cannot be stated too often that there are no fixed rules for writing propositions. The authors of these must recognize the form best adapted to the work at hand. Often, the result will be a very special proposition such as we have just studied.

VERBAL ORDERS

Contractors dealing with regular customers frequently receive verbal orders. The work is installed and billed for without any formal order ever having been issued. However, in most other cases, it is advisable to use some precautions.

Upon receiving a verbal order, a contractor must return to his office and write a letter to the buyer which reads appoximately as follows:

> We are proceeding in accordance with your verbal instructions
> to install the electrical work in your plant, located at . . .

The letter includes a description of the work as required. It may be sufficient to say "in accordance with plans and specifications," or if there are no plans and specifications, it may be necessary to write a lengthy description.

In addition to recording the price and description of the work, the letter forestalls the possibility of cancellation. Some time may pass before the work can get started. In the meantime, the buyer may decide to start shopping if no letter has been received. Upon receiving the letter, he will consider the matter definitely settled.

The contractor has much to gain and nothing to lose by writing letters acknowledging the receipt of verbal orders.

CHAPTER 14

What Type of Organization Do You Need?

Do you know what type of organization you need? The type of organization to be formed is a phase of starting business that must be studied carefully.

Much of our study and many of our examples have been confined to what we chose to call a "one-man business." Advantages as well as the limitations of this type of business will be covered as we study the types of organizations generally used in electrical-construction work.

There are three types of organization, or business, adapted to electrical contracting:

1. Sole owner and proprietorship
2. *a.* Partnership
   b. Limited partnership
3. Corporation

SOLE OWNER AND PROPRIETORSHIP

In the sole-owner and proprietorship type of organization, there is but one executive and he owns and has complete control of the business. This type of organization has its advantages as well as disadvantages. We shall first study the former.

Advantages

1. One man has complete control of the business.
2. The proprietor is not obliged to cater to the wishes of others.
3. Expenses can be kept at a minimum.
4. Tax problems are a minimum.
5. The owner is not responsible for debts created by others.

A business with one capable executive in control of all operations does not suffer the hazards of a company with several heads that are likely to start working at odds with one another. It also has the advantages of having operations proceed without being warped to comply with the wishes of co-owners.

We have studied the necessity of keeping down expenses while getting a business started. The individual ownership has only one executive salary, and office space can be limited. Minimum taxes are also an item to help keep down expenses.

As we study other types of organizations, we shall better appreciate some of the advantages of the individual ownership. It also has its disadvantages as shown by the following list:

Disadvantages

1. There is no relief from entire responsibility.
2. Capital is limited.
3. Credit is limited.
4. In case of sickness, there is no one to take over.
5. In case of death, there is no one to carry on.

The individual operator can never relax his hold on the business. He is often unable to get away for much-needed vacations. This affects not only his vacations but those of his family as well. In addition to being tied down, the sole owner does not have the advantage of the good counsel that is often supplied by capable co-owners.

Naturally, the amount of capital that the individual can raise by supplying cash or by borrowing is less than that of a group of two or more. Hence, operations may be greatly limited owing to lack of ready money or credit.

In electrical contracting, proprietors do not view sickness and death with as much concern as in the case of other businesses. After the contractor has been in business for some time, he should have a good foreman who is capable of carrying on the outside work for a while. His bookkeeper should be able to look after the office temporarily. However, if there were a sizable contract to be closed, a large project to be figured, or an important project starting, sickness could be a serious matter.

In case the owner dies, the business would more than likely have to be disposed of in some way. It is not easy for an estate to get a suitable man to carry on an electrical contracting business. If a man is capable of running the business and has to take all the responsibilities for its successful operation, he prefers to build up his own organization.

PARTNERSHIP

When contemplating a partnership form of organization, one must study many things besides those ordinarily listed as advantages and disadvantages of that type of firm. He must study himself to learn whether or not he has the right temperament for a partnership. If he is honest with himself, he may find that he lacks many of the necessary qualifications required for one who has to work so closely with others.

He must also appraise the contemplated partner. One may be a very good friend and provide good company socially but prove to be undesirable as a business partner.

We have often seen cases where two men working for the same contracting firm became disgruntled with the way the business was carried on. They got together regularly and put the boss "on the pan." Finally they decided that since the boss refused to be straightened out, they must start their own partnership. After being in business together for a while, they found out that they did not like each other any better than they did their former boss.

Among the advantages of the partnership form of business are the following:

Advantages
1. It can be strengthened by logical division of responsibilities.
2. More capital is available.
3. Credit is better.
4. In case of sickness or death of one partner, there is someone to carry on.

All the foregoing listed advantages are greatly dependent on the temperament, financial status, training, and natural abilities of the partners. Much can be revealed by discussing a combination that we have observed working well.

Let us call our partners *A* and *B*. *A* has been trained in the office. He not only understands the management phase of the business but is a good engineer, estimates well, and understands actual construction work as well. He manages the office, estimates the work, closes the contracts, buys the materials, and looks after the stockroom and billing. He is an all-around electrical contracting man.

A must be an honest man and deal fairly with his partner. He must also be able to meet the public and inspire confidence in his firm. Securing remunerative business depends on A's ability to estimate accurately and sell well.

Partner B has come up through the ranks as a mechanic. He has an average education, can superintend work, can select the better mechanics, is resourceful, and can deal with the customers in the field. B is essentially an outside man and directs all the installation work. He must also be honest and trust A.

A is thoroughly familiar with B's work and can lend a hand if necessary. However, B knows little about management and practically nothing about estimating. B should study estimating and learn something about A's other duties so that if outside work is light and estimating heavy, he can shift his activities.

The initiative must be taken by A, and B must be satisfied to be more or less the subordinate partner. With proper cooperation, A and B develop a combination that works effectively and they both enjoy the profits of a good business.

The operating capital and credit of a partnership depend on how much each wishes to put into the business and their personal financial status. However, we know that it would be greater than that of either individual.

With a partnership, chances of the business being carried on in case of sickness or death are much better than those of an individual ownership. In case of sickness, the ailing partner may not be able to go to the office but may be well enough to confer with the co-owner.

We are all inclined to be selfish and unfair in playing the "give-and-take" game of life. In a partnership, each partner must be prepared to give more than he thinks he should and be satisfied to accept less than he thinks he is rightfully entitled to.

In the discussion of partners A and B, we have already touched on many items that could breed difficulty in a partnership form of business. Let us list the disadvantages of such an organization and study them more in detail.

Disadvantages

1. Executive salaries high
2. Business too much dependent on disposition of partners

3. Difficulty in agreeing on the division of responsibilities
4. Difficulty in agreeing on salaries
5. Difficulty in agreeing on division of profits and expense allowances
6. Possibility of suspicion and jealousy creeping in
7. Each partner subject to debts created by the other

Some of the listings of disadvantages read like duplications but, when discussed, will be found to refer to different conditions.

Partners' salaries are bound to be a burden when the business is getting started. The partners may not allow themselves any pay, but their time is still a cost of operating the business. How long both salaries continue to be an overhead burden depends on how the partners spend their time. If conditions permit, one partner may start working with the tools as soon as a contract is signed and the job opens up. In this way his salary is a direct job expense instead of an overhead item. He will continue with the tools until there is enough work to employ him full time as superintendent. As superintendent, his time will continue to represent an item of job cost instead of overhead.

If both partners are white-collar men, the overhead will be excessive until there is enough business to require two men in the office. In most cases this is a year or more.

It is not necessary to spend much time on the hazards of bad temper, as that is an item with which we are all familiar. The hazards of quick tempers are universal and have created much greater havoc than the mere breaking up of a partnership.

When all is going well, there is not such a great possibility of disagreements regarding responsibilities, but when jobs begin to show a loss, trouble may start. B blames A and says he did not get enough money for the job. A blames B and says the job is not being run properly. Both are under a strain and likely to start an argument with the slightest provocation.

Disagreement on salaries, profits, and expense allowances stems from the same cause; it has to do with net gains. It, too, is more likely to occur when business is not good.

Before the business is started, the partners think they have all arranged so there will be no trouble about expenses, but as time goes on, things happen. A spends the firm's money freely for club dues and entertaining "prospective customers." B insists on buying a new car before one is needed and charges it to the business. So many things can happen to cause discord between partners.

As time goes on, B may get the idea that A is tampering with the books or that he is using the firm's money to buy things for his personal use. *B* may also become jealous of *A* because he is always out in front. *A* represents the firm at association meetings, goes to conventions, and has a great deal of time away from the office while *B* "slaves."

A overdoes his entertaining while *B* stays on the job and works to try and make the business pay. Again, we have observed cases where *A* was carrying the load and, to an extent, *B* was just so much dead weight.

Responsibility for partnership debts is a serious phase of partnerships. Any contracts signed or debts created are the responsibility of both partners regardless of which is the author. If the business fails and one partner cannot pay his share of the debts, the other is liable for the difference.

We may not have observed many instances where partnership debts proved to be a really serious matter, but the hazard is always there. If *A* signs a contract that results in a disastrous loss to the business, *B*'s home or anything he owns is in jeopardy until all debts are paid.

The reader may consider our picture of partnerships as dismal, but one must view a business from all angles and face the facts before entering it.

LIMITED PARTNERSHIPS

Some states have statutes providing for "limited partnerships." A limited partnership is for one wishing to put money in a business without taking an active part in the operations and without being liable for debts beyond the extent of the money invested. Although a limited partner is not generally liable for debts to an amount greater than he has invested in the partnership, he must be thoroughly familiar with the limitations placed on limited partners. Otherwise he may lose his immunity by being too active in the business.

A is a general partner and invests $20,000 in the business. *B* is a limited partner and invests $1,000 in the partnership. *B,* providing he maintains his immunity, is not liable for more than $1,000. However, *A* is liable for all debts.

A limited partnership must be authorized by the state in which the business is to operate. Anyone wishing to establish such a business connection must be sure that he is familiar with the state laws regarding it.

CORPORATION

A corporation is a legal organization in and of itself. It is a body with the legal right to act as an individual. Through its officers, it buys, sells, enters into contracts, and carries on business in much the same manner as an individual.

Before a business can operate as a corporation, a state charter must be obtained from the secretary of state in which the business is to operate. The first step is to file a certificate of incorporation. The following is, in part, information provided by the certificate:

1. Name of the corporation
2. Purpose—to provide electrical installations, to sell appliances and motors, etc.
3. Number of shares of stock to be issued
4. Location of the business
5. Number and names of directors
6. Names and duties of officers
7. Address of officers
8. Other pertinent items regarding the business and proposed operations

One intending to file a certificate of incorporation must be familiar with the laws of the state in which the business is to operate. Usually two or more go together to form a corporation. However, an individual may wish to incorporate his business and does so by creating officers and issuing stock certificates. His business has officers and directors and issues stock, but his management is actually that of an individual complying with laws of incorporation.

For the electrical contracting business, the following can be listed as advantages of being incorporated:

1. Limited liabilities for debts
2. Sales value
3. Attract investors (limited)

One owning stock in a corporation is liable for the corporation debts only to the extent of the money invested in that corporation. In case the company gets into financial difficulty, the stockholder knows that his home and other holdings are not in jeopardy.

In the electrical contracting business, being incorporated often has a sales advantage. All other things being equal, a corporation enjoys higher regard from architects, engineers, and buyers than the business operating as an individual or partnership. It is more likely to be stable, hence offers a better guarantee of satisfactory completion of contracts.

When we speak of attracting investors, we do not mean the investing public in general. Rather, we are thinking of existing stockholders and persons intimately acquainted with the corporation.

As a business continues to operate, its opportunities to secure better and larger contracts may increase. To take advantage of these opportunities, it is necessary to have more capital. A corporation would prove more inviting to investors than an individual or partnership form of business.

In the business world, taxes are regarded as the principal disadvantage of being incorporated. However, in the electrical contracting business, that is not a serious matter for some time. It takes time to build up a business that can pay dividends as well as remunerative salaries.

Corporation taxes are principally, if not solely, based on profits. To learn the effect of these taxes, two cases can be studied:

Case 1—the individual with his business incorporated
Case 2—the large corporation with stockholders outside the working organization

Case 1

For Case 1, we can use the "one-man" business that we have been studying and say it has been incorporated. We allowed $8,400 for the proprietor's salary, which was pointed out as being much too low. The amount could be increased to three times as much without being excessive. As business improved, salaries would be increased, stock added to, and facilities improved. It would be a long time before there would be dividends subject to corporate taxes.

Case 2

To share profits with the stockholders, officers of the corporation declare dividends. The dividends are subject to two taxes. They are taxed as profits of the corporation and, after being issued as dividends, are taxed as income of the individual stockholders.

Being incorporated involves some additional cost for bookkeeping, but compared with over-all cost of operating, the amount is not great.

ORGANIZATION A SERIOUS STEP

Selecting the type of organization is a serious step in starting a business. For the majority of cases it is found advisable to use the individual ownership or partnership form of organization when entering the electrical contracting business. After six months or a year, if the business is to be continued, it can be incorporated. If the business is not to be continued, the losses will not be so great as they would if the business had been incorporated.

CHAPTER 15

Can You Collect?

Are you familiar with the various methods of collecting for electrical installations? And do you know that the cases are limited where the contractor can just send a statement for the amount due him and expect to collect? For the most part, collections for electrical work involve considerable detail.

Six types of contracts can be listed, each of which involves some variation in billing. We shall study the following:

    I. Small fixed-price contracts
   II. Small cost-plus contracts
  III. Time and material orders with no stipulated markups
  IV. Large fixed-price contracts to be billed as work progresses
   V. Large cost-plus contracts to be billed as work progresses
  VI. Fixed-fee contracts

I. THE SMALL FIXED-PRICE CONTRACT

When the small fixed-price contract is completed, the contractor sends a statement for the amount of the contract. If a certificate of inspection is required, it should accompany the bill.

This is the simplest form of billing.

II. THE SMALL COST-PLUS CONTRACT

The small cost-plus contract, like the small fixed-price contract, is billed after the work is completed, and only one billing is required.

However, the cost-plus contract requires a breakdown billing with all material and labor hours listed.

Figure 15-1 illustrates the *method* of billing for a small cost-plus contract. The numbers in the left-hand column are for reference purposes and are listed along with the following discussion:

1. The contractor gives a statement of the total amount at the top of the sheet so the buyer can at once see the total amount of the bill. The buyer does not have the experience we all have with automobile repair shops. We look at a bill and think it is reasonable only to learn that there are charges listed all over the sheet with the total amount tucked away in one corner.

2. Being a cost-plus contract, it requires a quantity, base-cost, and extension listing for each item of material. Markups are added later.

3. One must watch his pricing carefully. Underpriced items go unnoticed, but one overpriced item may start trouble.

4. An item like the "ring hanger assembly" involves several parts such as rod, socket eye, ring, etc. Several listings are saved by the assembly pricing. Perchance one feels that it is good business to impress the buyer by having long listings of material; each part can be listed separately.

5. Cartage, tools, and insurances are listed along with material, and a common markup is applied to all. Both tools and insurances are incidental labor expenses but do not rate such high markups as labor.

6. One should not be careless about taking out city inspection if it is required. In the first place, it is the right thing to do, and in the second place, the owner likes to know his work is being inspected.

7. Eight per cent has been used for overhead markup on material. The material curve, Fig. 11-1, shows a higher markup for a contract of this size. There are several reasons why a contractor may feel justified in reducing the percentage for a project of this type.

 a. It is a select job with a quick turnover.
 b. Neither estimating nor engineering is required.
 c. It is a cost plus job and no risk is involved.
 d. Most of the material is high priced.
 e. A regular client rates a better markup than the occasional buyer.

8. The reasons advanced for a low markup on material for overhead apply to the markup for profit.

9. It is practice among electrical contractors to bill the mechanics' time as "hours labor." William T. Stuart, editor of *Electrical Construction and Maintenance,* contends that, instead, the billing should

MOSER ELECTRIC COMPANY

2020 West First St.   Arch Rock, Illinois

Clow and Clow
1040 Main Street
Arch Rock, Ill.

(1) For installing new power feeder to grinding room -- $1,546.00

    Itemized costs as follows:
(2) Power distribution Panel -------------------------- $ 225.00
(3) 110 ft. 3" galv. Conduit @ 130.00/c ---------------- 143.00
    3 3" galv. els. and cplgs. @ 7.30 ea. -------------- 21.90
    360 ft. 500,000 cm. RH-RW cable @ 1650.00/m ------ 594.00
    1 12"x16"x6" pullbox ---------------------------- 9.00
    15 3/8" beam clamps @ 20.00/c -------------------- 3.00
(4) 15 3" ring hangar assemblies @ 80.00/c ------------- 12.00
    4 3" insul. bushings @ 60.00/c ------------------- 2.40
    8 3" locknuts @ 30.00/c -------------------------- 2.40
    3 400A. fuses @ 3.00 ea. ------------------------- 9.00
    1 tape, solder and misc. ------------------------- 10.00
(5) Cartage -- 15.00
    Tools - 3% of payroll (224.00) ------------------- 6.72
    Insurances - 12% of payroll (224.00) ------------- 26.88
(6) City inspection -----------------------------------   5.00
                                                          1085.30
(7)   Overhead 8% --------------- 86.85
                                                          1172.15
(8)   Return 5% ----------------- 58.60   1230.75
(9) 64 Hrs. labor @ 3.50 ------------------------------ 224.00
(10)   Overhead 28% -------------- 62.72
                                                            286.72
(11)   Return 10% ---------------- 28.67   315.39

    Total cost of installation ----------------------- 1546.14

Fig. 15-1. A form of billing, No. 1 (see also Figs. 15-2 and 15-3). Unit costs are given, and markups to suit particular project are added as percentages of costs.

read "Hours—mechanics' time." There is much to justify his contention because the mechanic has many duties that do not involve labor.

The mechanic spends much of his time studying plans, laying out work, and studying job conditions. Such work is not labor. Again, the mechanic connects control wiring, adjusts equipment, and tests installations. Strictly speaking, such work is not labor.

The contractor has nothing to lose and may gain by using hours time instead of hours labor. Hours time may be a better tool for selling.

All the mechanics' time is listed at the base rate of journeymen electricians. The job is not large enough to require a foreman.

10. The labor overhead of 28% is also lower than that shown on the labor curve, Fig. 11-1, for a project of this size. In addition to some of the reasons given for low overhead on material, there is the advantage of having a job that is easy to man. It is easier for mechanics to adapt themselves to this kind of work than it is to commercial and multistory projects.

11. Ten per cent is added for return on labor. This is more or less standard. Contractors seldom add more and are not justified in using less. Any concessions due the customer are made in the reduced overhead markups.

III. THE TIME AND MATERIAL CONTRACT—
NO STIPULATED MARKUPS

The time and material contract is a form of cost-plus contract, but the contract does not stipulate the amount to be added to base costs to obtain a selling price. Such contracts are received from regular clients or new clients who have confidence in the contractor employed.

The form of billing for a time and material contract depends on the size of the project, discretion of the contractor, or wishes of the buyer. Small projects may be billed as shown by Figs. 15-1, 15-2, and 15-3. Large projects may have similar methods of billing but will be billed as the work progresses.

All other things being equal, the buyer giving a time and material order is entitled to better prices than the man who calls for competitive bids.

IV. LARGE FIXED-PRICE CONTRACTS—
TO BE BILLED AS WORK PROGRESSES

There is too much money involved in a large fixed-price contract for a contractor to wait until the work is completed before receiving any payments. Arrangements are made for him to make draws as the

MOSER ELECTRIC COMPANY

2020 West First St.   Arch Rock, Illinois

Clow and Clow

1040 Main Street

Arch Rock, Ill.

For installing new power feeder to grinding room - $1,550.00

Costs as follows:

1	Power distributuion panel --------------------	$ 260.00
110 ft.	3" galv conduit @ 148.00/c --------------	162.00
3	3" galv. elbows @ 8.25 ea. ------------------	24.75
360 ft.	500,000 cm. RH-RW cable @ 1870.00/m ------	673.20
1	12"x16"x6" pull box --------------------------	10.25
15	3/8" beam clamps @ 22.50/c ------------------	3.38
15	3" ring hanger assemblies @ 90.00/c ----------	13.50
4	3" insul. bushings @ 68.00/c ----------------	2.72
8	3" locknuts @ 34.00/c -----------------------	2.72
3	400A. fuses @ 3.40 ea. ----------------------	10.20
1	tape, solder and miscl. ---------------------	10.50
	Cartage -------------------------------------	17.00
	Tools ---------------------------------------	7.60
	Insurances ----------------------------------	30.50
	City inspection -----------------------------	5.70
64 hrs. mechanics time @ 4.93 --------------------		315.52
	Total ---------	1550.34

FIG. 15-2. A form of billing, No. 2 (see also Figs. 15-1 and 15-3). Unit prices are contractors' selling price. No percentages of markup are revealed.

MOSER ELECTRIC COMPANY

2020 West First St.   Arch Rock, Illinois

Clow and Clow   March 1, 19
1040 Main Street
Arch Rock, Ill.

For installing new power feeder to grinding room - $1,550.00
Costs as follows:

1	Power distributuion panel	$ 260.00
110 ft.	3" galv. conduit	162.00
3	3" galv. elbows	24.75
360 ft.	500,000 cm. RH-RW cable	673.20
1	12"x16"x6" pull box	10.25
15	3/8" beam clamps	3.38
15	3" ring hanger assemblies	13.50
4	3" insul. bushings	2.72
8	3" locknuts	2.72
3	400A. fuses	10.20
1	tape, solder and miscl.	10.50
	Cartage	17.00
	Tools	7.60
	Insurances	30.50
	City inspection	5.70
64 hrs.	mechanics time	315.52
	Total	1550.34

FIG. 15-3. A form of billing, No. 3 (see also Figs. 15-1 and 15-2). Neither unit costs nor percentages of markup are given.

work progresses. In outlining such arrangements, most contracts state that 85% of the amount due the contractor will be withheld until the work is completed and approved by the owner.

Breakdown of Contract

After the contract has been signed, the architect or some other representative of the owner requests an itemized breakdown of the contract costs. Figure 15-4 illustrates the type of breakdown requested. This division of costs is used by the owner's representative to check the value of work completed when the contractor makes a request for payment.

When the work is far enough along to justify a request for partial payment, the contractor prepares to make a draw. He studies the job and, after deciding which portions are far enough advanced to be included in his report, makes up an estimate of work completed. This estimate is sent in with the request for payment.

Estimate of Work Completed

Figure 15-5 shows an estimate prepared for use in connection with a request for partial payment on a contract. The divisions are listed together with their complete installation cost, the percentage completed, and the dollar value of the completed portion.

The first item in Fig. 15-5 is the main switchboard. The total cost (to the buyer) is $4,400 installed, and the installation is listed as 80% completed. As indicated in the listing, it is in place but not connected. The purchase price of a switchboard usually represents 80 to 90% of the total cost installed.

It will be noted that round figures are used throughout. The estimates are only approximate values, and it would add nothing to the value of report to go into odd figures.

The contractor lists the installed work at full value. This must be done because by the time he receives his money, considerably more of the installation will be completed. On a $43,000 project, the payroll would amount to $150 to $200 per day. Assume that it was $150 and that the material was an equal amount. That would be $300 per working day. If 2 weeks elapsed between the time the contractor sent in his application for payment and the time he received his money, with a 5-workday week, he would have an additional $3,000 spent on the job.

Amounts Withheld and Previous Draws

In Fig. 15-5 the total value of work completed was estimated to be $18,470. There are two deductions from this amount: (1) the 15%

BELMONT BUILDING, 320 West Belten Street

Divisions of Electrical Contract

Main Switchboard	$ 4,400.00
Power panels and cabinets	850.00
Lighting panels and cabinets	1,600.00
Lighting branch wiring - roughing in	7,200.00
Lighting branch wiring - trim	1,800.00
Feeders - power and light - conduit installation	3,500.00
Feeders - power and light - cable and trim	3,700.00
Telephone conduit and cabinets	1,200.00
Lighting fixtures and lamps	12,300.00
Power branch wiring conduit	1,200.00
Motors - wiring, starters, sw. and connect	2,750.00
Fire alarm - roughing	600.00
Fire alarm - equipment and wiring	1,200.00
Controls and misc.	800.00
Total contract	43,100.00

Fig. 15-4. Division of contract prepared for buyer.

February 1, 19

BELMONT BUILDING, 320 West Belten Street

Estimate of work completed to date

BRANCH OF WORK	TOTAL COST	PERCENT COMPLETED	VALUE OF WORK COMPLETED
Main Switchboard - in place	$4,400.00	80	$3,520.00
Power Panels and Cabinets	850.00	40	340.00
Light Panels and Cabinets	1,600.00	40	640.00
Light Bran. Wiring - rough	7,200.00	90	6,480.00
Light Bran. Wiring - trim	1,800.00	50	900.00
Feeders-power and lt.-condt.	3,500.00	90	3,150.00
Feeders-power and lt-cable	3,700.00	20	740.00
Telephone conduit and cab.	1,200.00	90	1,080.00
Power Br. W. - conduit	1,200.00	90	1,080.00
Fire Alarm conduit	600.00	90	540.00

Total value of work completed to date   18,470.00

Less 15% retained --------------------   2,770.00
                                          15,700.00

Amount of previous draws -----------   9,500.00

Balance due ------------------------   6,200.00

Amount of this request (see attached letter) $6,000.00

FIG. 15-5. Estimate of work completed to accompany request for payment.

which is to be withheld and (2) the $9,500 which the contractor had previously drawn.

Final Requests for Payment

In due time, after the work is completed, the contractor sends in his final statement. The final request for payment must have the certificate of inspection (if required) attached. Before the contractor receives full payment, he is required to issue a final waiver of lien. When receiving partial payments, the contractor issues a waiver for the amount received. Waivers of lien will be discussed in a later paragraph of this chapter.

V. LARGE COST-PLUS CONTRACTS— TO BE BILLED AS WORK PROGRESSES

The markups for material and labor are usually stipulated in the contract. There may be separate percentages for each or common markups. In some cases the labor is to be billed at so much per hour including all insurances, job costs, and markups for overhead and profit.

On a well-organized project, the owner has a representative who checks labor crews, approves timecards, checks material deliveries, and approves material sheets. Duplicate timecards and copies of material delivery sheets are given to the owner.

The contractor's billing is done from the approved timecards and material lists. The form of billing is similar to that shown by Fig. 15-1.

When materials are delivered directly to the job by the supply house, the contractor may elect to give duplicate copies of the invoices to the buyer and bill the total amount of each instead of listing each item of material. The pros and cons of supplying buyers with duplicate invoices have been previously discussed in Chap. 13.

The periods of billing may be stipulated in the contract or left to the discretion of the contractor.

VI. FIXED-FEE CONTRACTS

A fixed-fee contract is a form of cost-plus contract in which the markup is a fixed amount (lump sum). The estimated cost of a project may be $2,000,000, and the contractor agrees to manage the installation and furnish such services as are stipulated for the sum of $150,000. The cost may run more or less than the estimated $2,000,000, but the contractor's fee will remain $150,000. However, in a well-executed contract, there will be some limits placed on the total cost of the project and the length of time the contract is to remain in effect.

The fixed-fee form of contract is used more for the large general contract, embracing all trades. It was popular during World War II. The contractor's man would move in on the job and practically start from scratch. He would build an office, buy all necessary equipment, and man the job. Every item of expense was charged directly to the job. For his fee, the contractor did little more than build up an organization and supply some over-all supervision.

Since we are interested only in billing for the contract, merits and evils of methods used in letting fixed-fee contracts will not be discussed.

Plans for payment of the fixed-fee are outlined in the contract. At specified intervals, the contractor receives stated amounts until the total amount received has reached a certain point. A portion of the fee is withheld until the contract is completed.

VARIATIONS IN BILLING

As we have seen, causes for variation in methods of billing are not limited to the form of contract alone. Billing for the particular contract may offer many variations as was illustrated by Figs. 15-1 to 15-4.

In the illustrations, labor was all billed at the journeyman rate. On many projects (large ones in particular), billing is in accordance with the rating of the individual mechanic. This was illustrated in Fig. 13-4.

A successful electrical contractor, whose judgment is highly respected, once said, "You must study your customer and bill him in the manner that will get the best reaction."

WAIVERS OF LIEN

States have lien laws, commonly known as mechanics' lien laws, to ensure mechanics' wages. If the owner fails to pay for his work, a contractor can get a lien on the building. The building cannot be sold until the lien rights are satisfied. No owner wants a lien on his property.

A waiver of lien is a signed statement whereby the contractor releases all claims he has against the property. It usually serves as a combination receipt and forfeit of lien rights.

When contractors make draws on work before completion, a partial waiver is issued. After the contract is completed and paid for, a final waiver is issued. Usually, the final waiver accompanies the request for final payment.

A final waiver of lien is released up to the date issued. If the contractor must go back to the job and do additional work, his lien rights are reestablished.

WAIVER OF LIEN
MATERIALS OR LABOR

FORM No. 3-H
REVISED DEC. 1943

GEO E COLE & CO CHICAGO
LEGAL BLANKS

State of Illinois, } ss.
Cook COUNTY

October 1 19

TO ALL WHOM IT MAY CONCERN;

Whereas, We *the undersigned* Barr and Kissler Electric Company *have been employed by* George H. Mason *to furnish* The Electrical installation in accordance with plans and specifications *for the building known as* George H. Mason Office Building

Now, Therefore, Know Ye, *That* We *the undersigned, for and in consideration of* One *DOLLARS, and other good and valuable considerations, the receipt whereof is hereby acknowledged, do hereby waive and release any and all lien, or claim, or right of lien on said above described building and premises under the Statutes of the State of Illinois relating to Mechanics' Liens, on account of labor or materials, or both, furnished or which may be furnished by the undersigned to or on account of the said ... for said building or premises.*

Given *Under hand and seal this* 1st *day of* October *19*

Barr and Kissler Electric Co (SEAL)

Harry H. Barr (SEAL)
Pres.

FIG. 15-6. A final waiver of lien.

CHAPTER 16

Can You Conduct Labor-cost Studies?

Labor-cost studies are a must item for any well-operated electrical contracting business. To get the most out of such studies, they must be conducted in a methodical manner and the one conducting them must appreciate all the benefits to be gained. The following are some of the gains to be realized:

1. Establishment of new labor units
2. A better understanding and appreciation of existing tables of labor unit costs
3. Improved methods of installation
4. An appreciation of the effects of good management
5. A knowledge of the value of tools
6. A knowledge of the various factors that influence labor
7. Better distribution of mechanics
8. A foundation for establishing costs for the labor on new and special work

As we study the foregoing items, we see that they are more or less overlapping and in some cases interdependent. True, the primary purpose of cost studies is to learn how to estimate labor accurately, but to get the best results, all the items listed must be studied.

PROGRAM OF PROCEDURE

There are three general types of labor-cost studies, namely:

1. Breakdown studies (time and motion)
2. Job division
3. Completed project

Regardless of the type of study, there must be a definite program established. The first two methods require special attention when planning a program of procedure.

The contractor must make arrangements that will ensure accurate keeping of records. The records must be compiled in a manner that will reflect true detailed costs of the work. And arrangements must be made that will ensure completion of the studies.

Hundreds of cost studies have been started but never completed. Many were of short duration because arrangements for keeping records were not adequate. Others were never completed because the records, although voluminous, were faulty and not comprehensive. In other cases, the records were put aside until forgotten because the contractors did not want to spend the time required for careful analysis.

Methods of carrying on cost studies vary. In some cases a special clerk or cost-study man is sent to the project. He keeps all records, compiles his data, and puts the results in final form which reflects actual distribution of costs. This has proved to be the most satisfactory method.

Not all men are suited to cost-study work. A suitable man must have a keen insight into the significance of various labor operations. Seasoned estimators have made the greatest contributions to labor studies.

It is common practice for contractors to direct the foreman on the job to keep cost records. The method is in no way adapted to breakdown studies. For job-division studies, it has proved valuable in some instances and represented just so much wasted money in others. The failures are not necessarily a fault of the foreman.

As a rule, foremen have all they can do to keep the job running as it should, without keeping cost records. Some of them have no faculty for keeping such records. Besides, a man is not going to jeopardize his reputation by taking care of work that belongs to the office.

A foreman asked to keep job records should be given a special account number to charge his or any of his crew's time to when it is spent on cost-study work. Skimpy allowances are often responsible for poor results in labor studies.

BREAKDOWN STUDIES

These are studies of all labor occasioned by a particular piece of material or equipment. The following rule must be adhered to if true and representative labor units are to be established:

Labor must be charged to the item occasioning it and not necessarily to the item on which it is performed.

OUTLET BOXES
Breakdown Time Studies

Labor Item	Time in Minutes	
	Ceiling Outlet	Bracket Outlet
1. Roughing-in time:		
Distribute material	1	1
Layout—locate outlet	5	6.5
Knockout (2)	1	1
Cut and thread pipe (2½ in.)	8	8
Place box	2.5	4
Nipple down	..	6
Install cover	..	3
	17.5	29.5
2. Finishing—poling up	4.5	4.5
3. Prorata time:		
Study time	1	1
Preparation	2	3
Receive material and pick up scattered out, and miscellaneous	2.75	3.5
	5.75	7.5
4. Supervision	3.5	5.2
Totals	31.25	46.7
5. Final adjusted unit for part job	32	47
6. Units established for general use:		
Ceiling outlet (no cover)	35 min each—58 hr/100	
Bracket outlet	49 min each—82 hr/100	

FIG. 16-1. Breakdown study of outlet boxes. The "stop-watch" time is 17.5 min. The final selected unit is 35 min.

An elbow is installed. It is so located that a full length of conduit cannot be used. A piece must be cut and threaded. The labor for cutting and threading the conduit is charged to the elbow because the presence of the elbow occasioned the work.

Here we are going to study outlet-box labor. Again we shall see that work performed on conduit is not charged off as conduit labor. It is charged to the outlet box because in the absence of the box, it would not have been required.

Figure 16-1 lists the results compiled from studies made by the author and others of outlet-box labor. The studies were made for flat slab construction with reasonably uniform layout and normal working conditions.

INSTALLATION OF ELBOWS
A Breakdown Time Study

Labor Item		Time in Hours		
Conduit size, in.	1½	2	2½	3
Study time—ordering and checking	0.06	0.08	0.10	0.10
Unloading and delivering to storage	0.05	0.07	0.09	0.12
Moving from storage to work area	0.06	0.10	0.12	0.15
Measuring for installation	0.20	0.20	0.25	0.25
Cutting and threading conduit	0.30	0.45	0.60	0.75
Installation	0.30	0.50	0.85	1.00
Supervision and incidental labor—included				
Totals	0.97	1.40	2.01	2.37

The studies are for rigid conduit installed on a 12-ft. ceiling with machine threading. No fastening labor included.

Fig. 16-2. Breakdown study of elbows. The time for putting the elbow in place is approximately 40% of the total installation time.

A look at the values in Fig. 16-1 leads one to believe that clear-cut units were established for each operation. The original data would not be so convincing. Records for operations on separate parts of an individual project may vary as well as those for several installations. However, average values, some adjustments, and good judgment enable one to use such data to establish representative figures for future use.

Values resulting from breakdown studies, as a rule, must be readjusted when contemplated in connection with a "job-division" study. In turn, when estimating, the final unit must be adjusted to suit the particular job at hand.

The accompanying illustrations, together with the text, should convince the reader that establishing tables of representative labor units is not a simple operation. It involves more cost, care, and study than the novice can conceive.

A STUDY OF FIGURE 16-1

One not familiar with labor-cost studies will do well to study the listings given in Fig. 16-1. Enough material is provided in the listing to afford pages of text. Here the discussion must be limited.

Work Occasioned by the Outlet

Figure 16-1 portrays what has been stated in the text regarding work occasioned by a particular item. Cutting and threading of conduit and "poling up" are charged off as outlet-box labor. Neither the work on the conduit nor the poling up would have been required if there

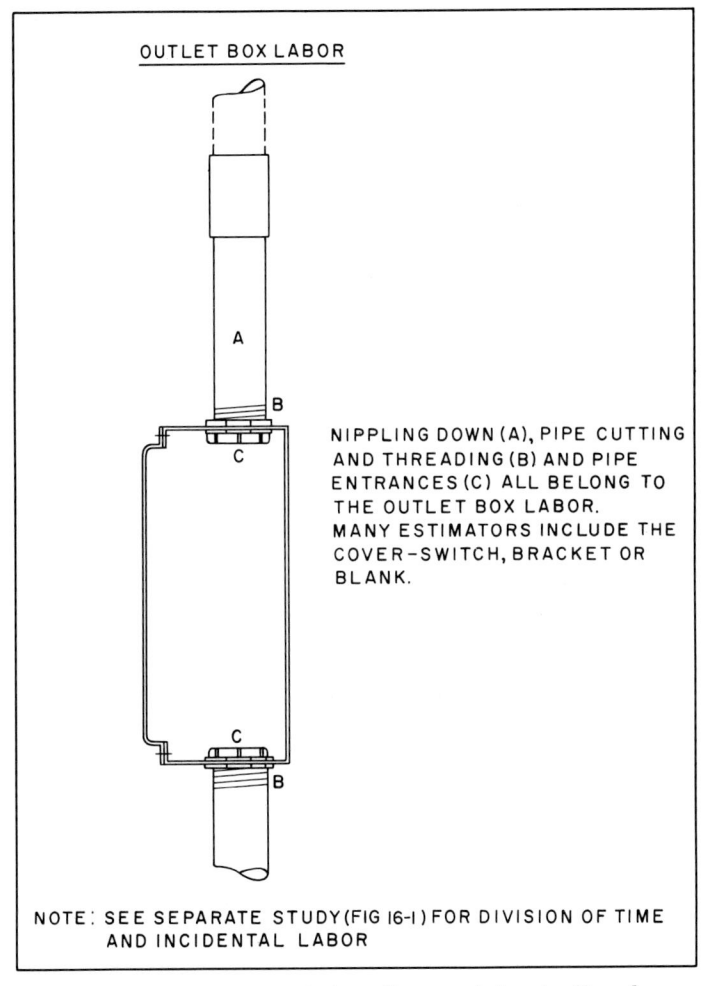

Fig. 16-3. Outlet box. Cutting and threading conduit, nippling down, and connecting are all part of the box labor.

had been no box. To have charged these items of labor to the conduit and wire and not to the box would have resulted in faulty labor units.

Incidental Labor

We hear much about so-called "stop-watch" estimators. The term refers to men who are frequently low bidders because they use labor

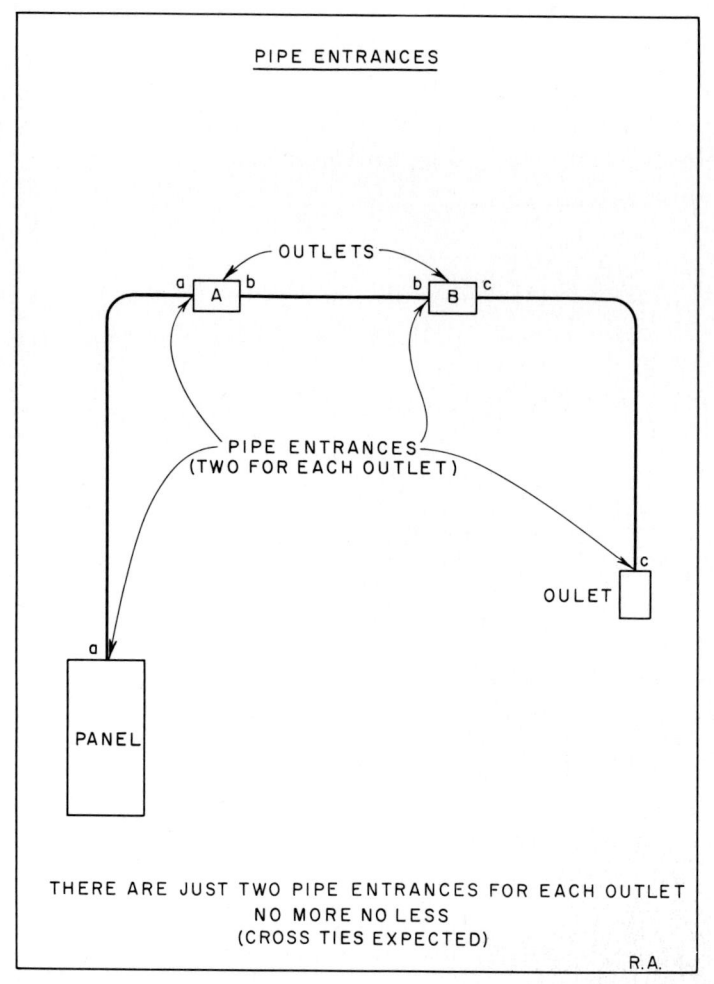

FIG. 16-4. Pipe entrances. There are two pipe entrances for each box, no more, no less.

costs established by incomplete studies. Usually the direct installation labor is accounted for but the incidental labor is overlooked.

Figure 16-1 shows the direct installation labor for an outlet box to be 17.5 min. A stop-watch estimator would consider his study complete at this point because the box was completely roughed in.

The 17.5 min represents the time spent on the deck, which is little more than 50% of the total time chargeable to the standard outlet-box unit. Study time, preparation time, receiving material, and all the other incidental labor must be included.

Supervision

Not all estimators include supervision in the base labor unit. Instead, a percentage of the total labor cost is added for supervision. There is nothing wrong with the method. However, the author and many others include an allowance in the base unit for two reasons:

1. Supervision costs vary. For example, the cost is greater for lighting branch wiring, special controls, and signals than it is for feeders and equipment.
2. If the units are used by beginners, supervision will not be overlooked.

Adjusted Unit for the Job

The total for the ceiling outlet was 31.25 min. The adjusted figure for the particular job is 32 min. Experience has taught us that when we adjust to get round figures, more accurate results are obtained by adjusting upward.

Units for General Use

The better estimators have long agreed that tables of standard labor units should use values approximately 10% greater than those established by careful studies. There are many reasons for this, but detailed discussion will be omitted here. However, one of the reasons should be explained in connection with this study. Better work results are generally realized on jobs where studies are being conducted.

Regardless of values used to make up tables, adjustments must be made to suit each project figured. Beginners need an extra 10% to compensate for parts of the take-off that have been slighted.

JOB-DIVISION STUDIES

Job-division studies, as the term indicates, are studies of divisions of work which embrace many items. Figures 16-9 and 16-10 reveal logical divisions for two different types of installations.

In Fig. 16-9, Parson Apartments, Division I, is for roughing in the branch-circuit work. Figure 16-11 lists the items embraced by this particular study. All outlet boxes, outlet box covers, fittings, and conduit for branch-circuit work are included.

CABLE INSTALLATION RECORD

JOB OR BLDG. THE TOWERS LOCATION 820 W. JEFFERSON

RECORD BY A.L. DATE 10/15/59

CABLE					HANDLING INSTALLATION EQUIPMENT						CABLE PULLING					MISC'L. LABOR & LOST TIME	TOTAL HOURS COMPL. JOB	HRS. PER 100 FT. OF CABLE				
SIZE	NO. PER COND.	LENGTH-FT. EACH LEG	LENGTH-FT. TOTAL	NO. OF ELS. IN RUN	TYPE OF EQUIPMENT AND SPECIAL RIGGING	LOCAT'N. OF SETUP	NO. MEN USED	HANDLING TIME-HRS. MOVE IN & SET UP	REMOVE & LOAD ON TRCK	TOTAL A	LOCT'N. REELS	NO. MEN USED	PULLING TIME IN HRS. MOVE IN & SET UP	CONN. TO CAB. & PULL	TOTAL B	C	A+B+C	ITEMS EQUIP. A	PULL'G. B	MISC'L. C	TOTAL	
500M	3	202±	USE 600	4	P.D. CAB PULLER	B	4	7	4	11	8TH	6	24	10	34	11	56	1.83	5.65	1.83	9.31	1ST PULL
500M	3	200	600	4	P.D. CAB PULLER	B	–	2ND	PULL		8TH	6	22	9	31	9	40	—	5.16	1.50	6.66	2ND PULL
No 1	3	99	USE 300	2	BLOCK & TAC.	1ST	1	10	5	1.50	1ST	2	1.50	3	4.50	1.20	7.20	0.50	1.50	.40	2.40	
3/0	3	95	285	2	B.& T. P.B. RT. TURN. 2 SET UPS.	1ST	3	2	1	3	1ST	3	4.50	5.40	9.90	2.75	15.65	1.05	3.46	0.97	5.48	
3/0	3	220	660	3	P.D. CAB PULLER	1ST	4	6 (MOVED FR. 8TH)	INCL.	6	B	4	6.00	6.50	12.50	4.00	22.50	0.91	1.90	0.61	3.42	
500M	3	90	270	2	CAB PULLER	1ST	4	3	4	7	B	5	7	5	12 –	3.50	22.50	2.58	4.95	1.30	8.33	
500M	3	90	270	2	CAB PULLER	1ST	–	2ND	PULL	–	B	5	7	5	12 –	3 –	15 –	—	4.45	1.11	5.56	2ND PULL

NOTES:
1. ELEV. & HOISTING SERVICE - GOOD
2. MECHANICS - AVERAGE
3. No SPCL. STORAGE & HANDLING

FIG. 16-5. Well-kept cable-installation records show the effect on labor of duplicate pulls, location of pulling equipment, length of pull, etc.

CABLE INSTALLATIONS[1]
DISTRIBUTION OF LABOR (IN HRS.) FOR THREE S.C., BR. & R.C., CABLES IN CONDUIT

SHEET 2A
SEE SHEET 1 FOR EXAMPLES

EXAMPLE & FT. OF CABLE PULLED	TYPE -SEE NOTE-[3]	PULLING EQUIP. MOVE IN & SET UP	REMOVE & LOAD	TOTAL A	MOVE IN & SET UP REELS	FISH	PREP. ENDS & BASKET	PULL TIME	TOTAL B	TOTAL PROD. LABOR A+B C	MISC. & NON PROD LABOR %	HRS.	TOTAL TIME-THE JOB C+D	HRS. PER 100 FT. CABLE SEE NOTE[2]
					— NO. 1 —									
I	300	B.&T.	1.00	0.50	1.50	1.50	0.75	0.75	1.50	4.50	6.00	20	1.20	7.20	2.40
II	300	F.T.	—	—	—	2.00	1.00	1.50	1.50	6.00	6.00	20	1.20	7.20	2.40
III[U]	300	F.T.	—	—	—	2.50	0.75	0.75	2.00	6.00	6.00	25	1.50	7.50	2.50
IV	600	C.P.	3.00	2.00	5.00	2.00	1.00	1.00	1.50	5.50	10.50	20	2.10	12.60	2.10
V	600	B.&T.	2.00	1.00	3.00	2.75	1.50	1.50	2.00	7.75	10.75	20	2.20	12.95	2.20
VI[U]	600	C.P.	4.00	2.00	6.00	5.00	1.50	1.00	2.50	10.00	16.00	25	4.00	20.00	3.35
VII	600	C.P.	4.00	2.00	6.00	5.50	2.00	1.00	3.50	12.00	18.00	25	4.50	22.50	3.75
					— NO. 000 —									
I	300	C.P.	3.00	2.00	5.00	3.00	1.00	1.00	1.50	6.50	11.50	20	2.30	13.80	4.60
II	300	B.&T.	2.00	1.00	3.00	4.50	1.30	2.00	2.00	9.80	12.80	20	2.56	15.36	5.12
III[U]	300	C.P.	4.00	2.50	6.50	5.00	1.25	1.00	1.50	8.75	15.25	25	3.81	19.06	6.30
IV	600	C.P.	3.50	2.00	5.50	5.00	1.50	1.00	3.00	10.50	16.00	20	3.20	19.20	3.20
V	600	C.P.	6.00	2.00	8.00	7.00	2.00	2.00	3.00	14.00	22.00	20	4.40	26.40	4.40
VI[U]	600	C.P.	5.00	3.00	8.00	9.00	2.00	2.00	4.00	17.00	25.00	25	6.25	31.25	5.20
VII	600	C.P.	4.50	2.75	7.25	12.00	3.00	2.00	4.00	21.00	28.25	25	7.06	35.06	5.85
					— 250 M C.M. —									
I	300	C.P.	3.50	2.00	5.50	4.00	1.00	1.00	1.60	7.60	13.10	20	2.60	15.70	5.20
II	300	C.P.	5.00	2.00	7.00	6.00	1.30	2.00	1.60	10.90	17.90	20	3.60	21.50	7.20
III[U]	300	C.P.	4.75	2.50	7.25	6.75	1.25	1.00	1.60	10.60	17.85	25	4.50	22.30	7.45
IV	600	C.P.	4.00	2.00	6.00	7.00	1.50	1.00	3.30	12.80	18.80	20	3.80	22.60	3.80
V	600	C.P.	7.00	2.00	9.00	9.00	2.00	2.00	3.30	16.30	25.30	20	5.10	30.40	5.10
VI[U]	600	C.P.	6.00	3.00	9.00	12.00	2.00	2.00	4.40	20.40	29.40	25	7.30	36.70	6.10
VII	600	C.P.	5.50	2.75	8.25	15.00	3.00	2.00	4.40	24.40	32.65	25	8.15	40.80	6.80
					— 500 M C.M. —									
I	300	C.P.	4.50	3.00	7.50	6.00	1.00	1.30	1.80	10.10	17.60	20	3.50	21.10	7.00
II	300	C.P.	7.50	3.00	10.50	9.00	1.30	2.60	1.80	14.70	25.20	20	5.05	30.25	10.00
III[U]	300	C.P.	6.00	3.75	9.75	10.00	1.25	1.30	1.80	14.35	24.10	25	6.00	30.10	10.00
IV	600	C.P.	5.25	3.00	8.25	10.00	1.50	1.30	3.60	16.40	24.65	20	5.00	29.65	4.95
V	600	C.P.	9.00	3.00	12.00	15.00	2.00	2.60	3.60	23.20	35.20	20	7.04	42.24	7.05
VI[U]	600	C.P.	7.50	4.50	12.00	20.00	2.00	2.60	4.80	29.40	41.40	25	10.35	51.75	8.65
VII	600	C.P.	6.75	4.00	10.75	24.00	2.00	2.60	4.80	33.40	44.15	25	11.05	55.20	9.20

[1] THESE STUDIES WERE MADE IN CONJUNCTION WITH THE COST DATA COMM. OF THE CHICAGO ELECTRICAL ESTIMATORS' ASS'N.
[2] STUDIES BASED ON:—
 1— 1938 LABOR CONDITIONS
 2— CONST. CREWS NOT EXCEEDING 20 MEN
 3— HOISTING TO BE FIG. SEPARATELY
[3] SYMBOLS—
 B.&T.— BLOCK AND TACKLE
 F.T.— FISH TAPE
 C.P.— CABLE PULLER
[U] — CABLE PULLED UP

ELECTRICAL CONTRACTORS' ASS'N. OF CITY OF CHICAGO
RWA

FIG. 16-6. Pulling requirements and labor hours vary greatly for cables of a given size.

Examples:

I—100 ft. straight run. II—100 ft. run with pull box at right angle turn. III—100 ft. run from fifth floor to basement. IV—200 ft. run, simple pull. Examples V, VI, and VII—See Fig. 16-6a.

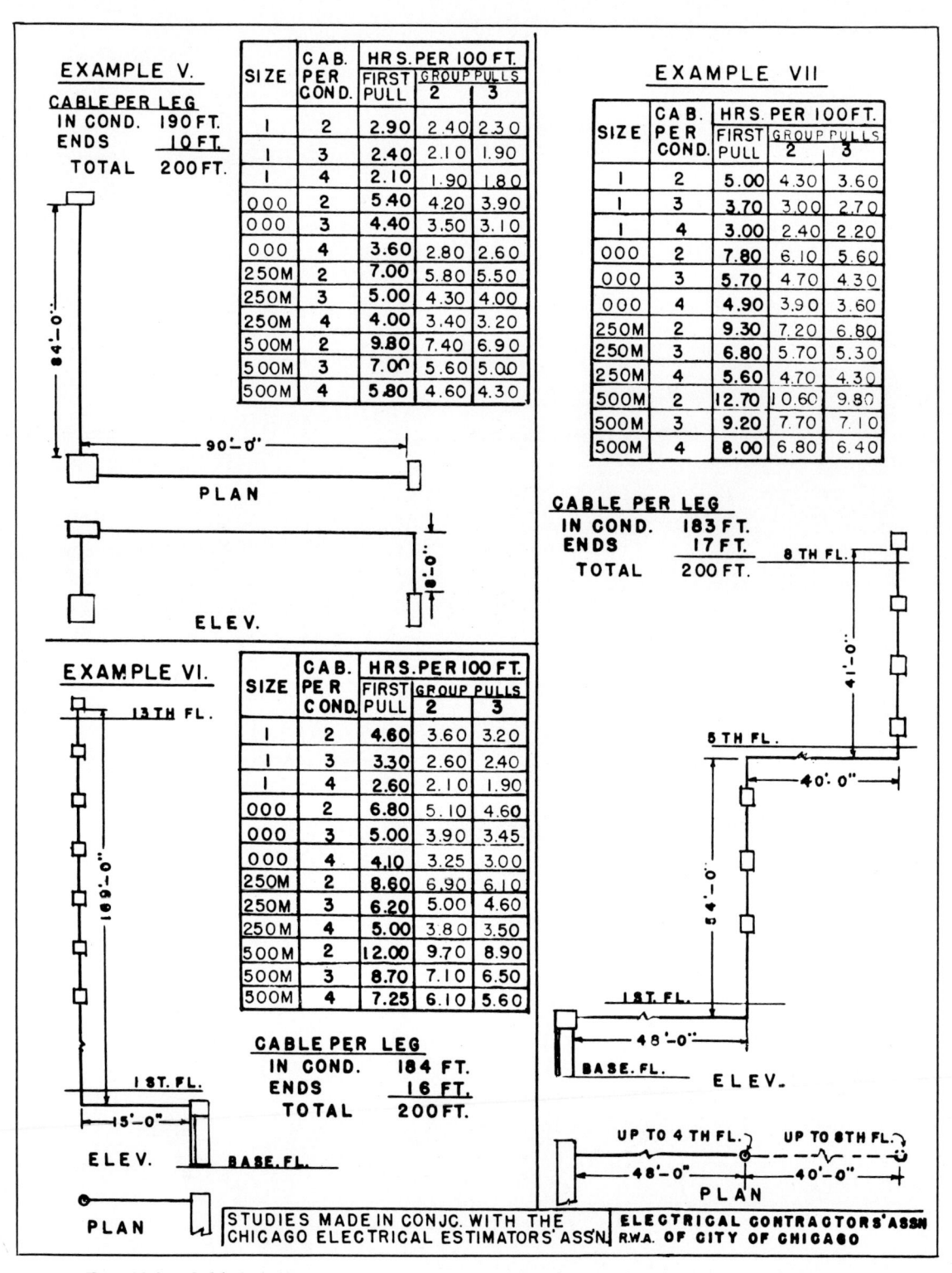

FIG. 16-6a. Cable-installation records. Layout and duplicate pulls are important factors.

CONDUIT INSTALLATION—RIGID STEEL.

Ⓝ APPROX. DISTR. OF LABOR FOR FEEDER COND. FASTENINGS & HANGERS NOT INCLUDED.

FOR 12 FT. MTG. HT. GROUP RUNS

SIZE	1/2"		3/4"		1"		1-1/4"		1-1/2"		2"		2-1/2"		3"		3-1/2"		4"	
WEIGHT—100 FT.	85		113		168		228		273		368		582		762		920		1089	
	HRS.	%	HRS.	%	HRS.	%	HRS.	%	HRS.	%	HRS.	%	HRS.	%	HRS.	%	HRS.	%	HRS.	%
STUDY TIME, ORDERING & CHECKING.	.20	5.	.22	4.4	.30	4.3	.35	4.0	.40	3.6	.45	3.0	.45	2.1	.45	1.8	.50	1.7	.60	1.5
UNLOAD & DELIVER TO STORAGE SPACE	.08	2.	.10	2.	.18	2.6	.30	3.4	.40	3.6	.50	3.3	.80	3.8	.90	3.6	1.10	3.7	1.40	3.5
MOVING FROM STORAGE TO POINT OF INSTALLATION	.08	2.	.10	2.	.18	2.6	.30	3.4	.40	3.6	.50	3.3	.80	3.8	.90	3.6	1.10	3.7	1.40	3.5
LAYING OUT RUNS & MEAS. FOR NIPPLES.	.75	18.7	.80	16	.80	11.4	.80	9.1	.90	8.2	1.00	6.7	1.10	5.2	1.10	4.4	1.20	4.0	1.50	3.7
SETTING UP AND HANDLING TOOLS.	.10	2.5	.10	2.	.14	2.	.30	3.4	.30	2.7	.40	2.7	1.00	4.8	1.00	4.0	1.25	4.2	1.50	3.7
CUTTING COND.—INCL. MEAS. 3 CUTS PER 100 FT.	.20	5.0	.23	4.6	.60	8.6	.70	8.	.80	7.3	1.20	8.0	1.80	8.6	2.25	9.0	2.70	9.0	3.50	8.7
THREADING (INCL. NECESS. CLEANING) 5 THREADS PER 100 FT.	.75	18.7	1.00	20.	1.30	21.4	1.95	22.2	2.40	21.8	3.50	23.3	5.20	24.8	6.60	26.4	8.00	26.6	10.75	27.0
INSTALLING CONDUIT (NOT INCLDG. EL'S. & OFFSETS)	1.00	25.0	1.25	25.0	1.55	22.	2.00	22.6	2.50	22.8	3.50	23.3	4.70	22.4	5.70	22.8	6.75	22.5	9.00	22.5
INSTALL EL'S. & CPLGS. 2 PER 100 FT.	.35	8.8	.50	10.	.60	8.6	.70	8.0	.80	7.3	1.25	8.4	1.60	7.6	2.00	8.0	2.50	8.3	3.25	8.1
OFFSETS—MAKING AND INSTALLING.	.34	8.5	.50	10.	.90	12.9	1.10	12.5	1.75	15.9	2.25	15.	3.00	14.3	3.50	14.0	4.20	14.0	6.00	15.0
CONNECTIONS AT P.B. OR CABINETS—2 PER 100 FT.	.15	3.8	.20	4.	.25	3.6	.30	3.4	.35	3.2	.45	3.0	.55	2.6	.60	2.4	.70	2.3	1.10	2.8
TOTALS	4.00	100.	5.00	100.	7.00	100.	8.80	100.	11.00	100.	15.00	100.	21.00	100.	25.00	100.	30.00	100.	40.00	100.

DIVISION OF TIME—HRS. PER 100 FT. & PERCENTAGES.

Ⓝ—COSTS VARY ACCORDING TO JOB CONDITIONS AND METHODS OF HANDLING WORK.

ELECTRICAL CONTRACTOR'S ASS'N OF CITY OF CHICAGO

FIG. 16-7. Conduit installation—group runs. Note that the item "Installing conduit" (putting conduit in place) represents only 22 to 25% of the total.

163

SINGLE RUNS

CONDUIT INSTALLATION–RIGID STEEL

—APPROX. DISTR. OF LABOR FOR FEEDER CONDUITS ON 12 FT. CONC. CEILING.—

DIVISION OF TIME (HRS. & PERCENTAGES) FOR 100 FT. OF CONDT. AND 2 ELBOWS.

SIZE OF CONDUIT	1-1/4"				1-1/2"			
WT. PER 100 FT. OF CONDT.	228 LBS.				273 LBS.			
	HOURS				HOURS			
LABOR OPERATIONS	CONDT. 100 FT. A	ELS. 2 REQ. B	TOTALS (A+B) HOURS	TOTALS (A+B) %	CONDT. 100 FT. A	ELS. 2 REQ. B	TOTALS (A+B) HOURS	TOTALS (A+B) %
STUDY TIME	.20	.10	.30	2.15	.23	.10	.33	1.90
ORDERING & CHECKING	.15	.05	.20	1.45	.5	.05	.20	1.15
UNLOADING & STORING	.35	.03	.38	2.70	.40	.04	.44	2.55
MOVING FROM STORAGE TO POINT OF INSTALLATION	.35	.03	.38	2.70	.40	.04	.44	2.55
LAYING-OUT RUNS & MEAS. FOR NIPPLES	.70	.15	.85	6.05	.75	.15	.90	5.15
SETTING UP & HANDLING TOOLS	.30	① —	.30	2.15	.40	① —	.40	2.30
CUTTING CONDT.-INCLDG. MEASURING. 3 CUTS/100 FT.	.20	.40	.60	4.30	27	.54	.81	4.65
CLEANING CONDT. (THREADS)	1.20	...	1.20	8.55	1.35	—	1.35	7.70
THREADING-5 PER 100 FT.	1.20	.80	2.00	14.25	1.35	.90	2.25	12.95
INSERTS (EXP. SLEEVES)	2.15	—	2.15	15.30	2.60	—	2.60	14.90
INSTALLING CONDUIT (NOT INCLDG. ELS. & OFFSETS)	2.40	—	2.40	17.20	3.00	-	3.00	17.20
INSTALLING ELBOWS 2 PER 100 FT.	—	.40	.40	2.85	—	.60	.60	3.45
MAKING & INSTALLING OFFSETS - 1 PER 100 FT.	.55	—	.55	3.95	.90	—	.90	5.15
CONNECTIONS AT PANEL BRDS. OR CABINETS - 2	.30	—	.30	2.15	.35	—	.35	2.00
PREPARATION TIME & MISCL. LABOR.-SEE NOTE ②	2.00	① —	2.00	14.25	2.85	① —	2.85	16.40
TOTALS	12.05	1.96	14.01	100	15.00	2.42	17.42	100
CONDUIT ONLY (LESS INSERTS)	9.90				12.40			
ONE ELBOW		98				1.21		

SIZE OF CONDUIT	2"				2-1/2"			
WT. PER 100 FT. OF CONDT.	368 LBS.				582 LBS.			
	HOURS				HOURS			
LABOR OPERATIONS	CONDT. 100 FT. A	ELS. 2 REQ. B	TOTALS (A+B) HOURS	TOTALS (A+B) %	CONDT. 100 FT. A	ELS. 2 REQ. B	TOTALS (A+B) HOURS	TOTALS (A+B) %
STUDY TIME	.25	.18	.43	2.00	.27	.20	.47	1.65
ORDERING & CHECKING	.15	.05	.20	1.00	.18	.06	.24	.85
UNLOADING & STORING	.50	.05	.55	2.60	.80	.07	.87	3.03
MOVING FROM STORAGE TO POINT OF INSTALLATION	.50	.05	.55	2.60	.80	.07	.87	3.03
LAYING-OUT RUNS & MEAS. FOR NIPPLES	.80	.20	1.00	4.80	1.10	.22	1.32	4.55
SETTING UP & HANDLING TOOLS	.60	① —	.60	2.90	1.50	① —	1.50	5.20
CUTTING CONDT.-INCLDG. MEASURING. 3 CUTS/100 FT.	.40	.80	1.20	5.70	.60	1.20	1.80	6.25
CLEANING CONDT. (THREADS)	1.50	—	1.50	7.20	2.25	—	2.25	7.75
THREADING-5 PER 100 FT.	1.50	1.00	2.50	12.00	2.25	1.50	3.75	13.00
INSERTS (EXP. SLEEVES)	2.60	—	2.60	12.50	2.60	—	2.60	9.00
INSTALLING CONDUIT (NOT INCLDG. ELS. & OFFSETS)	4.20	—	4.20	20.00	5.70	-	5.70	20.00
INSTALLING ELBOWS 2 PER 100 FT.	—	.80	.80	3.80	—	1.20	1.20	4.15
MAKING & INSTALLING OFFSETS - 1 PER 100 FT.	1.20	—	1.20	5.70	1.70	—	1.70	5.90
CONNECTIONS AT PANEL BRDS. OR CABINETS - 2	.50	—	.50	2.40	.55	—	.55	1.90
PREPARATION TIME & MISCL. LABOR.-SEE NOTE ②	3.10	① —	3.10	14.80	3.95	① —	3.95	13.75
TOTALS	17.80	3.13	20.93	100	24.25	4.52	28.77	100
CONDUIT ONLY (LESS INSERTS)	15.20				21.65			
ONE ELBOW		1.56				2.26		

SIZE OF CONDUIT	3"				3-1/2"			
WT. PER 100 FT. OF CONDT.	762 LBS.				920 LBS.			
	HOURS				HOURS			
LABOR OPERATIONS	CONDT. 100 FT. A	ELS. 2 REQ. B	TOTALS (A+B) HOURS	TOTALS (A+B) %	CONDT. 100 FT. A	ELS. 2 REQ. B	TOTALS (A+B) HOURS	TOTALS (A+B) %
STUDY TIME	.30	.20	.50	1.36	.35	.20	.55	1.30
ORDERING & CHECKING	.20	.07	.27	.75	.20	.10	.30	.70
UNLOADING & STORING	1.00	.10	1.10	3.00	1.20	.15	1.35	3.15
MOVING FROM STORAGE TO POINT OF INSTALLATION	1.00	.10	1.10	3.00	1.20	.15	1.35	3.15
LAYING-OUT RUNS & MEAS. FOR NIPPLES	1.20	.25	1.45	3.97	1.30	.30	1.60	3.75
SETTING UP & HANDLING TOOLS	1.75	① —	1.75	4.77	2.25	① —	2.25	5.25
CUTTING CONDT.-INCLDG. MEASURING. 3 CUTS/100 FT.	.90	1.80	2.70	7.35	1.10	2.20	3.30	7.60
CLEANING CONDT. (THREADS)	2.75	—	2.75	7.50	3.30	—	3.30	7.60
THREADING-5 PER 100 FT.	2.75	1.80	4.55	12.40	3.30	2.20	5.50	12.80
INSERTS (EXP. SLEEVES)	4.45	—	4.45	12.20	4.45	—	4.45	10.30
INSTALLING CONDUIT (NOT INCLDG. ELS. & OFFSETS)	6.90	--	6.90	18.80	8.25	—	8.25	19.10
INSTALLING ELBOWS 2 PER 100 FT.	—	1.50	1.50	4.10	—	1.80	1.80	4.20
MAKING & INSTALLING OFFSETS - 1 PER 100 FT.	2.00	—	2.00	5.45	2.35	—	2.35	5.45
CONNECTIONS AT PANEL BRDS. OR CABINETS - 2	.60	—	.60	1.65	.70	—	.70	1.65
PREPARATION TIME & MISCL. LABOR.-SEE NOTE ②	5.00	① —	5.00	13.70	6.00	① —	6.00	14.00
TOTALS	30.80	5.82	36.62	100	35.95	7.10	43.05	100
CONDUIT ONLY (LESS INSERTS)	26.35				31.50			
ONE ELBOW		2.91				3.55		

①—ELBOWS ARE NOT CHARGED WITH ANY LABOR WHICH WOULD BE NECESSARY FOR CONDUIT ONLY

②—"PREP TIME & MISCEL LABOR" REPRESENTS THE DIFFERENCE BETWEEN THE "STOPWATCH" TIME AND THE ACTUAL TIME REQUIRED FOR THE INSTALLATION. IT REPRESENTS DAILY "STARTING & PICKUP" TIME, UNAVOIDABLE INTERRUPTIONS AND DELAYS, ETC.

FIG. 16-8. Conduit installations—single runs. Time required for a single run of conduit is greater than it would be for the same conduit in a group run.

No attempt is made to segregate any particular system. All lighting, telephone, and signal roughing materials are included in the division.

Figure 16-10 gives the labor for a hospital. Division I includes conduit and boxes for the nurses' call system. If there had been small 120-volt fans and heating units, the branch-circuit conduit for these would have also been included.

Keeping records for roughing-in work is not usually difficult because little else is going on at the same time and practically all the mechanics' time can be charged to the one division. Later stages of the job are not so simple. Men may be working on feeders, panels, and trim work, all at the same time. In that case individual records must be kept for each division and incidental labor prorated.

A STUDY OF FIGURE 16-9

The results portrayed by Fig. 16-9 reveal experiences that are common when conducting labor-cost studies. The figures indicate that the first four studies involved no difficulty. However, Division V, Feeder Cables, was a "bust."

In the records there was no accounting for the handling of pulling equipment. The time for moving in and storing cable was included with incidental labor. The cable-pulling records were discarded.

Experiences like that for Division V are not unusual when foremen are keeping records for the first time. Besides, not being accustomed to the work, they have many other things on their minds.

Another item that looks out of line is Division VII, Bells and Annunciators. As indicated on the record sheet, the work was taken care of by an apprentice. Naturally, inexperience was reflected in the hours turned in. However, dollar-wise, the cost was little above normal.

Item XI, Motors and Starters, may startle the novice. However, the experienced estimator will not be surprised. In multistory buildings, with many motors on upper floors, it is not unusual for the electrical contractor to suffer the loss of a great deal of time because men have to await their turn to use the hoists or elevators. That is exactly what happened on the particular job. Much time was lost while trying to get motors up to the penthouse machinery room.

One may say, "Why doesn't an estimator allow for such hazards?"

Estimators do consider hazards, but if they figure all of them too liberally, they will be figuring themselves out of work. One has to assume that some of the bad breaks in one place will be offset by favorable conditions in another.

All things considered, the Parson (the original name has been

```
                        JOB STUDIES

Parson Apts.

100,000 Sq. Ft. - Six Fls, and Base.
Conc. Joist Construction
63 APTS.
```

DIVISION OF COSTS - In Hrs.

DIVISION		Estimated Hours	Job Cost Hours	% Change Abv. Av.	% Change Bel. Av.
I	Roughing in Branch Cts.	5264	4790		9
II	Pulling Br. Ct. Wire	1037	895		14
III	Feeder Conduits	444	518	17	
IV	Service Cables	98	90		9
V	Feeder Cables	371	160*		see text
VI	Switches and Receptacles	472	350		26
VII	Bells and Annunciator (appr.)	153	256	67	
VIII	Panels-Power	52	50		4
IX	Panels-Lighting	240	170		30
X	Meter Wiring (63 Apt.Met.)	260	190		27
XI	Motors and Starters	220	380	72	
XII	Apt. Meters - Set & Conn.	127	130	3	
XIII	Switchboard	160	140		12
XIV	Miscellaneous	300	150		50
		9,198	8,269		

* Did not include moving in and setting up cable pulling equipment

FIG. 16-9. Job-division studies. Studies provide multipliers for establishing correct labor units to fit the particular installation.

changed) Apartments provided a very valuable study. It was the first apartment building the contractor had signed up in a long time, and it was felt that current conditions warranted study. Future contracts for similar buildings benefited greatly by the study and experiences.

A STUDY OF FIGURE 16-10

The data in Fig. 16-10 were taken from one of the most satisfactory studies the author ever conducted. We had contracted for many hospitals, but this was the first one to be studied carefully. A well-trained foreman, with the aid of an apprentice, kept the records for the bulk of the job, and their data were turned in in excellent condition.

The time for roughing-in, Division I, was greatly in excess of the estimated time, but the general contractor was slow and his progress erratic. Besides, the weather was bad and all trades worked against odds. However, after the slabs were all poured, work moved nicely and the job rolled on to completion in good time. Again it was a case of bad breaks being offset by good ones.

Studies of other hospital projects indicated that the labor units used for Division I were substantial.

ADJUSTING LABOR UNITS

Before we enter into the methods for adjusting labor units, a word about the values used in Figs. 16-11 and 16-11a.

The quantities used are essentially the same as those required for the actual installation, except round figures have been substituted. The units have been uniformly increased to values somewhat above those experienced on the particular project.

The experienced units, owing to excellent working conditions, good mechanics, and careful management, were exceptional. To have used them might have been misleading. Published units, of necessity, must be near standard.

The author has more recent studies, but those in Figs. 16-11 and 16-11a were chosen because Division III, Fig. 16-11a, shows the influence of breakdown studies. It will be noted that the values assigned to some job units are much greater than those used in the estimate. The breakdown studies dictated such a change for all elbows.

In early days of electrical estimating, there was a tendency to prorate labor in proportion to the cost of the material. As a result, labor for boxes and elbows was underpriced and that for conduit was overpriced.

JOB STUDIES

Aurora Park Hospital

39,000 Sq. Ft. - 70 Beds
Six Fls. and Base.
Conc. Joist Const.

DIVISION OF COSTS - In Hours

DIVISION		Estimated Hours	Job Hours
I	Roughing of Branch Circuits	1611	1886
II	Feeder Conduits	370	315
III	Br. Ct. Wire (poling up not incl.)	467	415
IV.	Switches and Receptacles	122	72
V	Feeder and Service Cables	115	130
VI	Lighting Panels and Cabinets	132	120
VII	Power panels and cabinets	66	52
VIII	Meter Wiring	42	40
IX	Nurses' Call Equipment	131	92
X	House Phones (7 10 button)	35	38
XI	Doctors' Call Equip.	125	85
XII	Motors and Starters	160	184
XIII	Bell wiring	40	60
XIV	Cap Outlets	20	20
		3436	3509
	Foremans Time	830	636
	Total Hours	4266	4145

FIG. 16-10. Job-division studies. Unusual job conditions produce a wide gap between estimated hours and job hours.

PRICING SHEET

JOB OR BLDG. PARSON APTS.	LOCATION
BID TO 100,000 SQ. FT. 6 FLS. & BASE.	~~ADDRESS~~ CONC. JOIST CONST.
ARCHT. OR ENG'R. 63 APTS.	ADDRESS
PLANS MARKED FAST MOVING JOB - GOOD MECHANICS	SPEC. No. EST. No. 4810
EST. BY PRICED BY EXTENSIONS BY	CHECKED BY S.O.No. SHEET No. 3

MATERIAL	QUANTITY	MATERIAL UNIT	EXTENSION	LABOR UNIT	EXTENSION
DIVISION I					
ADJUSTED LABOR UNITS					
FOR ROUGHING IN BR. CTS.					
EST. COST — 5264					
JOB COST 4790					
ADJ. FACTOR 0.91					
			ESTIMATED LAB		ADJ JOB
			UNITS EXTEN		UNITS
			HRS		(slide rule calc.)
CEILING OUTLETS — IN CONC. JOIST	950	0.60	570 —		.55
" " SUS. CEIL.	100	.75	75 —		.68
4" SQ. BOXES & SW. COVRS.	2400	.65	1560 —		.59
4" " " & BRKT. "	900	.70	630 —		.64
4" EXTEN. RINGS & SW. COVERS	110	.25	27 50		.24
4 11/16" SQ. BOXES	5	1.00	5 —		.91
HANDY BOXES	30	.70	21 —		.64
GEM SW. BOXES	15	.70	10 50		.64
PLASTER RINGS — 4"	580	.10	58 —		.09
½" RIGID CONDT.	44600	4.50/C	2007 —		4.10/C
¾" RIGID CONDT.	3000	5.00/C	150 —		4.50/C
INCIDENTAL LAB. (NOT DISTRIBUTED)			160 —		14560
TOTAL			5264 —		

FIG. 16-11. Job-cost study of branch-circuit work.

PRICING SHEET

CHI. ELECT. EST. ASSOC. FORM 4

DATE _____ 19____

JOB OR BLDG. PARSON APTS. LOCATION _____
BID TO LABOR STUDIES ADDRESS _____
ARCHT. OR ENG'R. DIV. II & III ADDRESS _____
PLANS MARKED _____ SCALE _____ SPEC. No. _____ EST. No. 4810
EST. BY A PRICED BY _____ EXTENSIONS BY _____ CHECKED BY _____ S.O. No. _____ SHEET No. 4

MATERIAL	QUANTITY	MATERIAL UNIT	EXTENSION	LABOR UNIT	EXTENSION
DIVISION II — PULLING IN CIRCUIT WIRES —					
LIGHT & SIGNALS					
EST. TIME 1037 HRS. JOB TIME 895 HRS.					
895/1037 = 0.86 = ADJUSTMENT FIGURE					
		EST. UNIT	EST. HRS EXTEN.		ADJ. UNIT-HRS
No. 18 TYPE "R" WIRE	9000	5./M	45—		4.3/M
No. 16 " " "	18000	5./M	90—	JOB UNITS 86% OF EST.	4.3/M
No. 14 " " "	125000	7./M	875—		6./M
No. 10 " " "	1200	12./M	1440		10.3/M
No. 14 " " " (Long Runs)	2000	6.5/M	13	ADJ.	5.6/M
TOTAL			103740		
DIVISION III					
FEEDER, TEL. & SIG. CONDTS. (1" & UP)					
EST. TIME - 443.5 HRS.					
JOB TIME - 518. HRS.				UNITS NOT	
EXCESS LAB 17% - APPROX				INFLUENCED BY JOB RECORDS	
1" RIGID CONDUIT	950	.07	6650		.07
1¼" " "	1300	.09	117—	JOB UNITS	.10
1½" " "	370	.12	4440	FOR EACH SIZE	.12
2" " "	650	.15	9750		.15
3½" " "	70	.28	1960	ADJ.	.33
1" ELS. & COUPLINGS	10	.20	2—	UNIFORM	.40
1¼" " " "	50	.30	15—	NOTE:	.50
1½" " " "	20	.40	8—		.70
2" " " "	15	.60	9—		1.00
3½" " " "	3	1.50	450		3.00
MISCL. HGRS. & FASTENINGS	300	Av. .20	60—		.30
TOTAL			44350		
THIS WAS A COMB. "BREAK-DOWN" & JOB-DIVISION STUDY					
HENCE THE RADICAL INC. IN ELBOW UNITS.					

FIG. 16-11a. Job-cost study of branch-circuit wiring and feeder conduits. Unusual job conditions require special adjustment of units. Note Division III.

It has taken some contractors a long time to get reconciled to putting labor where it belongs. For that reason, some readers may find the outlet-box units higher than they have been accustomed to using. These same readers should also find the conduits priced below their standards. It must be remembered that labor must be charged to the item occasioning it. Hence some conduit labor must be charged to boxes and elbows.

Figure 16-11 shows the total estimated time for Division I to be 5,264 hr. The actual time used on the job was 4,790 hr, which was 91% of the estimated time.

It was assumed that the labor on all items benefited equally by the saving and that each labor unit was entitled to a 9% downward adjustment. The last column of Fig. 16-11 shows the adjusted labor units for the particular project. Several such studies must be made before standard estimating units can be established.

Figure 16-11a provides the adjusted job-cost units for wire pulling, Division II, and feeder conduits, Division III. The adjustment unit for Division II was 86%. As shown in the illustration, the estimated cost of pulling No. 18 wire was 5 hr per 1,000 ft. The adjusted job unit was 86% of this amount, which was 4.3 hr per 1,000 ft. Items for other wire sizes were reduced the same percentage.

There are times when details of the cost records indicate that not all items in the division should be subject to the same revision. Such was the case for materials listed in Division III. Some of the job units are the same as those used for estimating, whereas others are adjusted upward. If the job units had been applied to the quantities, the total extended hours would have been 518, which is the same as the actual job hours.

One can readily see that studies such as have been illustrated are a valuable asset. To keep up with changing times and conditions, contractors must conduct them regularly.

THE COMPLETED-JOB STUDY

Almost every completed fixed-price contract receives some attention. The attention ranges from a cursory examination of cost records to an extended study of details. Since the most attention is given to projects which show excess labor, they will be considered first.

Contracts That Lose Money

When contracts have excess labor or have lost money on both labor and material, there is a tendency to select one or two conditions that have been opposed to the work and assign all losses to those conditions. The causes selected may be the major items but by no means the only factors responsible for the losses.

A contractor contemplating losses on a completed project must consider the following:

1. Was the right foreman selected?
2. Did the foreman have suitable mechanics?
3. Was the job overmanned?
4. Were plans and shop drawings adequate?
5. Was progress of the general contract and other subtrades normal?
6. Were the proper tools supplied?
7. Was the delivery of materials well timed?
8. Were the materials adapted to the installation?
9. Were the working spaces normal or cluttered with materials and equipment?
10. Was there anything unusual about the type of project?
11. Did the mechanics have the proper support from management?
12. Was any particular branch of the work responsible for the excess labor?
13. Was each branch of the work estimated correctly?

A contractor must have a program to follow if he is to get a complete and true picture of the details as well as the job as a whole.

Too often men are selected because they have a good record without consideration being given to their adaptibility to the work at hand. A man may have a good reputation for certain types of work, but that does not mean that he will fit all conditions. This applies to both foremen and mechanics. A good mechanic may not be a good foreman, and a good conduit man may do very poor work when assigned to control wiring.

Other things besides a man's ability as a mechanic must be considered. His likes and dislikes, ability to learn new work quickly, habits, and ability to get along with others must all be considered.

Overmanning Jobs

A trouble akin to selecting the wrong men is that of overmanning work. This frequently happens and the most common causes are:

1. There is a temporary shortage of contracts and the contractor does not want to lay off good men.
2. The contractor gives way to the demands of the owner for more men on the job.
3. Management is careless.

Overmanning a job is serious and costly. Often the additional men can do very little to speed up completion. An installation can go just so fast, and any number of men assigned to the work will not affect the progress of the general construction and other trades which control the speed of the electrical contract.

Besides the direct cause for lost motion due to overmanning, there are indirect causes. The mechanics get the idea that they are working on a cost-plus contract and the contractor is trying to build up the man-hours. Again, the mechanics get the impression that the work has been figured liberally and can afford lost time. If mechanics think the employer is not worrying about the cost of the job, they will be lax.

"On-and-off-the-job" mechanics is another form of man-power supply that has often caused labor costs beyond those estimated. Large or medium-sized projects are used as reservoirs to supply men for meeting the fluctuating man-power demands of the smaller jobs. This imposes a costly burden on the job that has to take care of the "on-and-off" mechanics.

Every time a mechanic leaves or returns, it costs the job money. When he leaves, adjustments have to be made to fill the gap, and when he returns, adjustments must be made so there will be a place for him. These interruptions and adjustments take their toll in time lost. Besides, the men that come and go do not take a normal interest in the work.

Too often losses due to the effects of on-and-off-the-job mechanics are overlooked when trying to ferret out the causes for excess labor.

Material Deliveries

The seriousness of untimely delivery of materials must not be overlooked when conducting completed-job studies. In a previous chapter, we learned of the necessity of having the right materials delivered in the right quantities and at the right time.

It is hard to evaluate losses due to material deliveries; however, some allowance must be made to avoid overloading other items of cost.

CONTRACTS COMPLETED AT LESS THAN ESTIMATED COST

Completed contracts which show a marked saving in costs should be studied as thoroughly as those showing losses. Again the contractor must ask questions. This time regarding the causes for the saving:

1. Was it a marked saving on the purchased material?
2. Could the material saving have been anticipated when the work was estimated?

3. Was it the progress of the general construction work?
4. Was it because the mechanics were especially well adapted to the work?
5. Was it the support given the men in the field?
6. Was it the type of installation?
7. Was it a combination of many favorable factors?
8. Was it because the estimate was too liberal?

These and many other questions must be settled if the study is to be effective.

Completed costs below those estimated are a signal for close study. They may indicate that the contractor has not been operating effectively on other jobs or that he has been losing contracts because of faulty estimates.

Under the heading "Job-study Benefits," we shall learn why studies enable contractors to get more work with less estimating.

JOB-STUDY BENEFITS

The foregoing examples and text have acquainted the reader with many values of completed-job studies. A summary of all job benefits would include the following:

1. Management mistakes can be detected and corrected.
2. The values of mechanics and foremen for particular types of work can be appraised.
3. A better knowledge of how to select the better contracts is gained.
4. One can know the cost of procurement failures.
5. General contractors, engineers, and architects can be appraised and classified as to the desirability of their work.
6. Willingness of other trades to cooperate can be noted.
7. It is possible to get more work with less estimating by taking full advantage of the knowledge gained.

The estimator who is in position to anticipate all factors that will affect the cost of a project has a distinct advantage. He can carefully select the work to be figured and figure according to its merits. The results are less estimating and more work.

The completion of a cost study does not mean that the contractor can go out the next day and get more contracts at better prices. He must wait for an opportunity to use the knowledge gained.

The results of gains from studies are more or less intangible. However, we cannot deny that studies are a vital part of the electrical contracting industry. Many contractors do not realize it, but their better business operations are the result of information passed on to them by the more progressive operators.

Research reveals that great advancements have been made in electrical contracting in the last 40 years. Outstanding progress has been made in the following phases of the business:

1. Over-all management
2. Training of mechanics
3. Engineering and construction plans
4. Tools and construction equipment
5. Dealing with architects and engineers
6. Dealing with customers
7. Methods of bidding cost-plus work
8. Methods of bidding installation-only work
9. Billing customers
10. Estimating

To this list, many other phases could be added. All the advances are the result of research and job studies.

CHAPTER 17 *

Do You Understand Incidental Costs?

NONPRODUCTIVE LABOR A MISNOMER

Promiscuous use of the term *nonproductive labor* has led to much confusion in the electrical-construction industry. The term has been used loosely and carelessly. Electrical estimators frequently have long lists of items of productive labor under the heading "nonproductive." This careless use of what should be a well-understood term cheapens the contractor's services and hampers his progress.

Application of this term has been varied according to the wishes of the user. Contractors endeavoring to establish accurate labor costs at first used it to designate the unaccounted-for labor. They kept careful cost records of all operations, but when the job was completed, the total time expended was not all accounted for. In the beginning, just this unaccounted-for time was called "nonproductive labor." As time went on, a long list of items was included under this classification. Finally, it was decided by representatives of a national association to call *all labor not consumed in the actual installation of the materials* nonproductive.

This decision was made in an effort to devise some means of explaining to the uninformed contractor that certain portions of labor cannot be isolated and must be prorated to the major operations. Like many other efforts to enlighten men who are not able to create their own labor units, it boomeranged.

In over 25 years of estimating jobs of all types and sizes, I cannot

* Copied and reedited from *Electrical Construction and Maintenance*, 1951.

recall ever listing an item of labor expense that could be rightfully classed as nonproductive. Among my acquaintances are some top-notch estimators who have also avoided the use of the term. In the first place, on a well-managed job, the amount of nonproductive labor should be nil. In the second place, if there are expenses for labor that does not produce, they can be given a better classification.

Waiting for other trades and time lost due to procurement failures are two common causes of labor expense that produce nothing. These or any other similar items can be classed as *incidental* when one thinks it advisable. Otherwise they can be listed individually and called just what they represent.

When it is expedient to offer some approximate division between actual installation labor and miscellaneous labor, the terms *job labor* and *incidental labor* can be used.

It is far better to use the term incidental labor than NPL (nonproductive labor). NPL is a negative selling term and generally produces a poor psychological effect. Owners want to buy productive labor, not NPL, and it does not boost the morale of men to have their backbreaking work called nonproductive. *Besides, handling materials, setting up tools, and study time are all just as productive as putting conduit in place.*

An owner is contemplating a contract on which he will supply the material for the contractor to install. He will be more impressed with the gravity of the situation if told that procurement failures will cause useless loss of time than he will if told there will be NP labor.

The Labor Unit

The standard labor unit should include all "on-the-job" labor. Figure 17-1 shows several items of incidental labor, such as handling materials, study time, layout, and other pertinent operations all included in the labor unit for 3-in. conduit. It must be remembered that the figures given are for *standard estimating units*. They will have to be altered according to special hoisting requirements, offsets, difficult working conditions, and other unusual job conditions.

Every estimator has his own ideas as to how he wants to build up his labor units. However, the majority of estimators want to include all incidental on-the-job labor.

Miscellaneous Labor

You will note that the listing for 3-in. conduit in Fig. 17-1 has an allowance of 5.20 hr for *preparation* and *miscellaneous* labor. The 5.20 hr is about 19% of the total time. To reconcile this time, one

ESTIMATED DISTRIBUTION OF LABOR FOR 3-INCH CONDUIT AND 3-INCH ELBOW

3-Inch Conduit

Labor Operation	Hrs. per 100 ft	% of total
Study Time	.40	1.50
Ordering and Checking	.30	1.10
Unloading and Storing	1.00	3.65
Moving from storage to point of installation	1.00	3.65
Setting up and handling tools	1.75	6.35
Laying out runs & meas. for nipples	1.30	4.75
Cleaning threads	2.50	9.10
Cutting Cond't.- incl. measuring (els not incl.)	1.30	4.75
Threading cond't. 3 per 100 ft.) els not incl.)	3.00	10.90
Making offsets (one per 100 ft.)	2.25	8.25
Installing conduit	5.20	19.00
Preparation & miscellaneous labor (see note)		
	27.50	100.00

3-Inch Elbow

Labor Operation	Hours Each	% of total
Study time	.10	3.2
Ordering & checking	.05	1.6
Unloading & storing	.10	3.2
Moving from storage to point of installation	.10	3.2
Setting up tools	(See Cond't.)	
Laying out runs & meas. for nipples	.40	12.5
Cutting Condt. - incl. measuring	.50	15.5
Threading conduit	.80	24.8
Installing Elbow	.75	23.5
Preparation & Miscellaneous Labor	0.40	12.50
	3.20	100.00

Note - "Preparation and Miscellaneous Labor" includes daily get ready time - morning and noon, pick up time at night, and normal relief periods

Fig. 17-1. Standard labor unit should include all normal on-the-job labor. This breakdown of such a unit for 3-in. conduit includes several items of incidental labor, study time, handling of materials and tools, ordering and checking, and others.

must take into account that the time for each operation is what is known in the trade as "stop watch." Some place in the listing must provide allowance for daily preparation time—morning and noon, pickup time at night, relief periods, and the general job preparations.

On the jobs where the conduit is concealed in concrete, there is an item of labor expense for men watching it while the concrete is being poured. There may also be standby time waiting for or on other trades. *All these labor costs are an essential part of the installation, and if essential, they are productive.*

As presently used, the term "nonproductive labor" is, in effect, a misnomer in our industry. It can be replaced by terms that more closely define the specific labor items covered. It may take a lot of verbal prodding to get us out of the NPL rut, but I am sure that no one will lose anything if we never again hear the term nonproductive used in the discussion of the electrical contracting industry.

INCIDENTAL LABOR AND DIRECT JOB COSTS

In previous paragraphs the fallacy of classifying incidental labor as nonproductive was discussed. Assuming that the suggested classification of incidental labor is correct, there is still the problem of how it should be presented in the estimate. As we have the same problem regarding direct job costs, the two items of expense can be treated together.

Every office must have check lists as reminders for its estimators. A complete list of incidental job costs is something one rarely finds in contractors' offices. The contractor operating near home and on jobs of a similar nature, year in and year out, may get along with estimated percentages based on the nature and volume of the project. Nevertheless the list of expenses making up these percentages must be available.

Figure 17-2 gives a listing of some of the numerous items of incidental labor and other items of direct job expense that may be encountered. In checking these items with Walter Brand of the Newbery Electric Corp., Los Angeles, the author was advised that each estimator of that organization was provided with a reminder list for checking incidental expenses.

Mr. Brand also advises that projects in hot arid sections call for many expenses not normally encountered. The supply of ice water alone becomes quite an item.

The proper method of listing and charging incidental costs must be left to the estimator. An item of labor may be a definite part of

INCIDENTAL LABOR AND DIRECT JOB EXPENSES

EQUIPMENT AND GENERAL

Engineering
Drafting
Blue Printing

Tools—Consumed and Depreciated
Scaffolding
Trucks
Gas, Oil and Truck Supplies
Rental Equipment
Freight & Cartage

Field Office and Shop Buildings
Furniture & Equipment (Field Bldgs.)
Lavatory & Sanitary Facilities (Field Bldgs.)
Wiring Field Bldgs. (Mat.)
Wiring Temp.—on the Job (Mat.)
Light & Power (Bills for Current)
Job Telephone
Heating Field Bldgs.
Ice Water (Appreciable item in some Sections)

Travel Expenses:—
   Air, R.R. and Auto
Room and Board
Special Exp. (For men away from home)
Rental for Storage Space (Off the Job)
Painting Materials
Special Hoisting Charges
Surveying
Barricades & Lanterns
Prorata Charges

Licenses
Inspections
Legal Expenses
Association Dues
Interest on Pay-Roll
Reserve for Contingencies & Guarantee

LABOR

Supervision
Travel Time
Erecting & Removing Scaffolding
Special Hoisting & Material Handling
Wiring Field Bldgs. (Labor)
Wiring— Tem. on Job (Labor)

Job Engineer
Job Draftsman
Time Keeper
Watchman
Job Clerk

Painting (Labor)
Testing
Erecting Barricades
Traffic Control
Lost Time (Anticipated) due to—
  a. Bad Weather
  b. Procurement Failures
  c. Job Interruptions
Clean-up

BONDS AND INSURANCES

Performance Bond
Special Guarantees
Property Damage
Property Damage—Installation Floater Policy
Fire Insurance
Builders Risk
Workmen's Compensation
Public Liability
IBEW Benefit Funds
Federal Taxes
State Taxes

Notes:
(1) Items of cost should be listed on the estimate sheets where they are best suited. Painting and testing may be identified with a particular phase of the work, whereas engineering, tools, field buildings and similar expenses are listed wth the items of Incidental Direct Job Expenses.
(2) Incidental labor expenses such as watchman and insurances may be listed in the material column of the Direct Job expense sheet. They are items of expense that do not require the same high markup as the mechanics payroll.

FIG. 17-2. Check list of incidental labor and direct job expenses like this should be provided to each estimator to assure inclusion of all such items.

a particular branch of the work on one job whereas it may be a general expense on another. Testing for a signal system should be charged off with other items of that system. Over-all testing specified to be carried on in the presence of the owner's representative is general.

An item of expense may be for labor, but for markup purposes it will be listed in the material column. Labor insurance is such an item. There are two reasons why this should be subject to markups designed for material rather than for labor. First, the cost of supplying insurance is more in line with the cost of supplying material. Second, one is inviting trouble if he submits estimates or renders bills with too high a markup on insurance.

On a medium-sized contract for a complete installation the corresponding overhead markups for material and labor would be approximately 10 and 35%, respectively. The smart contractor just will not try to collect 35% for overhead on insurance.

Incidentals Are Not Overhead

In spite of all that has been written about keeping direct job expense out of overhead, it is still in order to mention the importance of having all such items of cost clearly identified. Without going into details, three reasons for this can be advanced: (1) It is imperative for accurate estimating; (2) direct job costs are easier to sell than overhead; (3) it is accepted practice in the construction industry.

Figures 17-3 and 17-4 provide listings and spaces for contractors to use in estimating job costs and overhead for a particular job or business. The former is for *direct job costs* and is designed for use in connection with the individual job. The latter is generally used in connection with *over-all business operations and costs*.

A study of the items listed in Figs. 17-3 and 17-4 establishes the difference (normal) between direct job costs and overhead. As this discussion is not on overhead, Fig. 17-4 is presented for comparison only.

Attention is again called to the fact that the *items of incidental labor are usually productive*. In the long list of items shown in Fig. 17-2, the only really nonproductive item is lost time. If travel time is necessary, it is in a way productive.

Substantiating Costs

Previous paragraphs have been concerned with making provisions for including all costs. Now methods must be pointed out for presenting and substantiating such incidental charges.

Again, it must be noted that the application of separate markups

A—DIRECT JOB COSTS

MATERIAL—SERVICE

Items of Expense	Est cost %	Dollars
1—Material by contractor Engineering & layout		
Estimating		
Selecting & purchasing		
Follow up & Coord. delivery		
Cartage & special delivery		
Storage facilities—Field		
Field tel.—Prorata		
Special reports—defense reg.		
Association dues		
II—Material by Owner		
Engineering & layout		
Assisting owner's pur. agent		
Follow up & coord. delivery		
Storage facilities		
Procurement failures (see note 4)		
TOTAL		

LABOR SUPPORT

Items of Expense	Est. cost %	Dollars
Tools—consumed & depreciated		
Cartage—tools		
Field office & shop bldgs.		
Engineering & layout		
Estimating		
Drafting		
Field tel.—prorata		
Supervision		
Blue printing		
Heat & light—field offices & shop		
Power wiring & current—field shop		
Travel expense		
Insurances		
Inspection		
Prorata charges		
Association dues		
Interest on payroll		
Procurement failures Material by owner—Lab. disruptions (N.-4)		
Res. for conting. & guarantee		
TOTAL		

B—OVERHEAD & ADMINISTRATIVE EXPENSE

Items of Operating Costs Not Identified with Any Particular Project

Items of Expense	Check fig.	Est. Cost %	Dollars
Administrative salaries	5.00		
Engineering & est.—missionary work.	0.80		
Bookkeeping—gen.	1.04		
Spec. bookkeeping—ins. & taxes	0.52		
Steno. & tel. oper.	0.45		
Store rm. attendant	0.67		
Utility boy	0.37		
Rent—office	0.70		
Rent—store rm.	0.20		
Light	0.12		
Telephone	0.35		
Office equip. & furniture	0.25		
Stationery, est. forms. & misc. supplies	0.29		
Postage	0.13		
Taxes & legal exp.	0.25		
Advertising	0.20		
Insurance on equip. & misc. exp.	0.65		
Research & time studies			
Adj. factor-for vol. fluctuation	0.60		
TOTAL			

1—Percentages compiled from surveys of compl. instlns.—all mat. & lab. by electrical contr.

FIG. 17-3. Special forms for figuring direct job costs are particularly useful for selling purposes.

FIG. 17-4. Cost items which make up general overhead on $600,000 volume.

183

PRICING SHEET

CHI. ELECT. EST. ASSOC. FORM 4	DATE NOV. 1	19 ___
JOB OR BLDG. HALL PLATING	LOCATION STOCKTON HTS.	
BID TO DO.	ADDRESS DO.	
ARCHT. OR ENG'R. ✓	ADDRESS ✓	
PLANS MARKED SEE T-1	SCALE ✓ SPEC. NO. ✓	EST. NO. 872-D
EST. BY PRICED BY EXTENSIONS BY	CHECKED BY S.O.NO.	SHEET NO.

SHEET NO.	MATERIAL	QUANTITY	MATERIAL UNIT	EXTENSION	LABOR UNIT HRS.	EXTENSION
	~ SUMMARY & BID SHEET ~					
1	CONN. AT PRES. SW. BRD.			15—	24	60—
2	DISTR. PANELS			250—	14	35—
3	FEEDERS			350—	60	150—
4	POWER BR. WIRING			240—	110	275—
5	MOTORS — SET & CONN.			30—	60	150—
				885—		670—
	INCIDENTALS		2%	18—	5%	34—
				903—		704—
	INSPECTION — 8M. — 90 H.P.			15—		
	INSURANCES — LAB. (10% OF 704)			70—		
	OTHER DIRECT JOB COSTS		4%	36—	12%	84—
				1024—		788—
	OVERHEAD & ADMIN. EXP.		10%	102—	35%	275—
				1126—		1063—
	SERVICE RETURN		—%		—%	
	SELL PRICES					

FIG. 17-5. Percentage method of handling incidental and direct job costs for average-size projects is shown at bottom of this bid summary sheet. Note incidentals, insurance, and inspection listed as separate costs with insurance in material column.

Do You Understand Incidental Costs?

for material and labor may seem revolutionary to the reader who has followed the traditional custom of using a common markup. However, *it is the only method that will provide accurate costs.*

Figure 17-5 illustrates what is known as *a percentage markup* for direct job expense. It is in reality a combination of individual items (inspection and insurance) and percentages. Assuming that the percentages used are basic, the method is very good for run-of-the-mine small jobs and medium-sized jobs (up to approximately $10,000) which

	Material	Labor
DIRECT-JOB-COST PERCENTAGE BREAKDOWN		
Estimating	1.15	1.35
Engineering and drafting (including field engineering)	0.80	2.88
Blueprinting	...	0.10
Supervision	...	2.60
Tools—consumed and depreciated	...	3.50
Selecting materials and expediting delivery	1.40	...
Cartage and special deliveries	0.20	0.30
Interest on payroll	...	0.50
Prorata charges	...	0.05
Travel expenses, job, stationery, and miscellaneous	0.45	0.40
Reserve for contingencies and guarantee	0.10	0.30
Totals	4.10	11.98

FIG. 17-6.

have no special expense such as hoisting, job buildings, time-keeper, and similar special requirements.

"Incidental expenses" may be included with each system (distribution panels, feeders, etc.) or on the summary sheet as shown in Fig. 17-5. When listing incidental expenses with each system, one must be careful not to have his estimate loaded with too much for minor items. Naturally, the percentages included on the individual or summary sheets will vary with the extent to which minor items of cost are listed on the estimating sheets.

In Fig. 17-5, insurance, a labor expense, has been listed in the material column and is subject to 10% for overhead. The reasons for this have often been pointed out in previous chapters.

For "Other direct job costs," Fig. 17-5 provides 4 and 12% for material and labor, respectively. The breakdown listing for these figures is approximately as shown in Fig. 17-6.

List Job Expenses

Figure 17-7 shows an estimating sheet with a listing of direct job expenses. The data at the top of the sheet, giving the estimated mandays, men required, and duration, serve as a guide for estimating costs for timekeeper, telephone, and bills for electricity. The listing embraces several items of expense not required for the type of project depicted by Fig. 17-5. Job office, shop buildings, wiring for buildings, and utility bills are special. Such items cannot be set up as a percentage of labor or material. Their costs must be lump sums estimated for the particular project.

The total estimated costs shown in Fig. 17-7 are transferred to a bid sheet as shown by Fig. 17-8. An additional item of direct job expense—insurance and benefit fund—was not figured until the labor listed in Fig. 17-7 was added to the other estimated labor. This term of expense can be included on the sheet with other incidental expenses if the estimator elects. The important thing is to have a *consistent practice* so the item will not be overlooked.

Among estimators there are three items of expense known as the "three I's." They are incidentals, insurances, and inspection. These are standard items in most localities, and it is well to have a uniform method of listing and checking them.

Substantiate Charges

Regardless of how direct job expenses are figured, the contractor must always stand ready to substantiate his figures. In lump-sum bidding a contractor must hew close to the line. To get the most remunerative contracts he must be sure that all charges are correct.

A large percentage of the work done by electrical contractors consists of some form of cost-plus contract. In negotiating with buyers, one must substantiate his charges if he is to get a remunerative price. Regular customers often want fixed or upset prices. A contractor must have an estimate which clearly depicts all costs. He must also stand ready with figures and data to substantiate any charges made.

Billing Labor

Figure 17-9 presents a condensed listing of operating-cost items designed to substantiate billing rates for labor. Normally, all the first seven items listed under "Direct labor expenses" are included in one listing and labeled "Insurance and association dues." The individual items, as shown here, must be available for the customer's information.

CHI. ELECT. EST. ASSOC. FORM 4		PRICING SHEET			DATE		19

JOB OR BLDG. **MAXWELL CENTER BLDG.** LOCATION
BID TO SEE T-1 ADDRESS **SEE T-1**
ARCHT. OR ENG'R. ADDRESS
PLANS MARKED SCALE SPEC. No. EST. No.
EST. BY **A. H.** PRICED BY EXTENSIONS BY CHECKED BY S.O.No. SHEET No. **16**

MATERIAL	QUANTITY		MATERIAL UNIT	EXTENSION	LABOR UNIT	EXTENSION
— INCIDENTAL & DIRECT JOB COSTS —						
{APPROX 720 MAN-DAYS 6 MEN (AV.) 24 WKS.						
{SEE PRELIM. LISTING						
ENG. & DRAFTING				525 —		
BLUE PRINTING				50 —		
SUPERVISION (INCL. IN LAB. UNITS)						
TOOLS-CONSUMED & DEPR (3% OF LAB.)				450 —		
CARTAGE				100 —		
HOISTING — ALLOW FOR SPCL. CHARGES				250 —		
JOB OFFICE & SHOP BLDGS. — DEPR.				125 —		325 —
WIRING JOB BUILDINGS				50 —		150 —
ELECT. — EST. BILLS				60 —		
TEL. — 6 MO. AT $10.00				60 —		
WATCHMAN — BY G.C.						
TIMEKEEPER & UTIL. MAN	WKS	20	70.—	1400 —		
ALLOW FOR PRORATA CHARGES				150 —		
TRAVEL EXP. (R.R. & OFFICE TO JOB)				180 —		
INSPECTION LT. {130 GTS. 1200 UNITS PR. {33 M. OUT. 120 H.P. EQ				175 —		
INTR. ON PAYROLL (1%)				150 —		
{INCIDENTALS & RES. FOR CONTING & GUAR.				550 —		600 —
{APPROX. 2.5% MAT., 4% LAB.						
TOTALS				4275 —		1075

FIG. 17-7. Detailed listing is used to determine incidental and direct job costs on large projects where special requirements (job office, watchman, etc.) preclude use of percentage method. Total costs are then added to bid summary sheet.

PRICING SHEET

CHI. ELECT. EST. ASSOC. FORM 4

DATE **NOV. 10** 19—

JOB OR BLDG. **MAXWELL CENTER** LOCATION **1220 SO. PIERS ST.**
BID TO **MEYER & BOWEN** ADDRESS **420 W. ARLINGTON**
ARCHT. OR ENG'R **DO** ADDRESS **DO**
PLANS MARKED SCALE SPEC. No. EST. No.
EST. BY **A. H.** PRICED BY EXTENSIONS BY CHECKED BY S.O.No. SHEET No. **17**

SHEET NO.	MATERIAL	QUANTITY	MATERIAL UNIT	EXTENSION	LABOR UNIT	EXTENSION
	~ BID SHEET ~					
15	SUMMARY — SEE 15			21450 —		15200 —
16	INCIDENTAL & DIR. JOB COSTS — SEE 16			4275 —		1075 —
				25725 —		16275 —
	INS. & BENEFIT FUND (10% OF LAB.)			1628 —		—
				27353 —		16275 —
	OVERHEAD & ADMIN. EXP.		6%	1641 —	21%	3418 —
				28994 —		19693 —
	SERVICE RETURN		—%		—%	
	SELL PRICE					

FIG. 17-8. Bid summary sheet has incidental and direct job-cost figures, developed in Fig. 17-7, added to total material and labor cost of estimate. Insurance (in material column for low markup) is added to total before overhead is applied.

Aside from the "Insurance and association dues," there are only two other items of direct job expense listed: "Engineering" (estimating and engineering) and "Tools." In billing it is intended that "Supervision," "Cartage," "Job offices," and any "Special expenses," will be included as individual charges on the cost sheet.

The overhead expense listing in Fig. 17-9 is very much condensed and may not always satisfy buyers' demands. The contractor must be able to produce more detailed listings, similar to those provided in a previous example. Upon close study of this figure, the reader will note that the overhead is not applied to the direct job costs. To compensate for this irregularity, the overhead percentage has been increased to approximately 2% above normal.

The figures are for the average cost of a $300,000 volume. With 100 separate contracts, the average would be $3,000. Curves show the overhead for a $3,000 contract to be approximately 21%. A figure of 23% was used in the study.

The differential of 2% can be justified as follows: Applying 10% (material overhead) to the $0.45 job cost produces $0.045. As it is, 2% was applied to the hourly rate of $2.50 and the result was $0.05. There is a $0.005 error, but the figures are only for the purpose of providing approximate costs to substantiate the correct billing.

Figure 17-10 provides a listing of approximate costs (contractor's) for labor per hour, where the base rate for mechanics is $2.35 per hour. The insurance rates used in preparing the values were the same as those used in Fig. 17-9, and other percentages were practically the same. The percentage used in Fig. 17-10 can be used to substantiate charges for any base rate within 10% of $2.35, as the figures are approximate. The method of estimating hourly charges will be the same for any rate.

SUPPLY DETAILED COSTS *

Costs similar to those shown in Figs. 17-9 and 17-10 are not in themselves sufficient. Breakdowns of the individual items must be available. In Fig. 17-9, 3% is included for tool costs. The contractor must be prepared to give a listing of the required tools and the cost per month for each. He may even go further and provide a comparison between the charges and those authorized by the government for comparable tools.

Charges for engineering and other items must be justified by more

---

* For additional analysis of incidental costs see *Electrical Estimating,* McGraw-Hill Book Company, Inc., New York, 1956.

Analysis of Contractor's Labor Costs for Journeymen Electricians

Average Cost of Items to be Added to a $2.50 Hourly Wage Scale (see Notes) Based on an Annual Volume (Cost) of $300,000.00 with a "60/40" Ratio.

Per Cent of Payroll	Direct Labor Expense	Cost per Hour
3.00	I. B. E. W. benefit fund (Electrical Insurance Foundation)	0.0750
1.00	Association dues	0.0250
2.80	State unemployment contribution	0.0700
0.30	Federal unemployment contribution	0.0075
1.00	Federal old-age social security	0.0250
2.50	Workman's compensation insurance	0.0625
1.50	Public liability, property damage, occupational disease, etc.	0.0375
3.00	Estimating and engineering	0.0750
3.00	Tools and construction equipment (consumed and depreciated)	0.0750
18.10		0.4525

Per Cent of Payroll	Overhead Expense—See Note 4	Cost per Hour
2.00	Rents—office and storeroom (including light and heat)	0.050
2.00	Postage, furniture, telephone, and stationery	0.050
1.00	Taxes, licenses, legal, and miscellaneous expenses	0.025
7.00	Salaries for office employees (including storeroom attendant)	0.175
2.00	Travel and sales expenses (automobile, railroad, hotel, etc.)	0.050
8.00	Administrative salaries	0.200
1.00	Adjustment factor (3 bad years in 10)	0.025
23.00		0.575

Base pay—Class A journeymen electricians			2.50
Direct labor expense		18%	0.45
Overhead expense		23%	0.575
Estimated net labor cost per hour			3.525

Notes:

1. To determine costs for apprentices, foremen, and superintendents, add 41% to wage scale paid.
2. Supervision, cartage, and special delivery costs, to be added as direct job expense.
3. For installation-only (material by others) projects, add 4.5% to direct job expense and 5% to overhead.
4. Percentages of costs vary according to type of job and the individual contractor's experience.
5. No allowance has been made in the above tables for service return (anticipated profit).

Fig. 17-9. Contractors find it much easier to get a legitimate hourly rate for labor if they can furnish the buyers with a detailed list of costs involved.

CONTRACTORS COST (APPROXIMATELY) PER HOUR
FOR SUPPLYING THE SERVICES OF A CLASS A
JOURNEYMAN ELECTRICIAN RECEIVING $2.35 PER HOUR

(See Notes)

Size of project (man-days)	Labor, markup for job costs and overhead			Labor, total cost per hour	
	Job costs, %	Over-head, %	Total, %	Complete installations 60% material 40% labor	Installation-only projects (add 5% to complete installation cost)
5–10	20	55	75	4.10	4.30
10–20	19.6	45.4	65	3.87	4.08
20–40	19.4	38.6	58	3.71	3.90
40–60	19.2	35.3	54.5	3.64	3.82
60–75	19	33	52	3.57	3.75
75–100	18.6	30	48.6	3.49	3.66
100–125	18.4	28	46.4	3.44	3.62
125–175	18.2	26.3	44.5	3.40	3.57
175–200	18	25.2	43.2	3.37	3.54
200–250	18	24.5	42.5	3.35	3.52
250–300	18	23.5	41.5	3.32	3.49
300–400	18	23	41	3.31	3.48
400–500	18	22	40	3.29	3.45
500–600	18	21	39	3.27	3.43
600–800	18	20	38	3.25	3.41
800–1000	18	18.5	36.5	3.21	3.37
1000–1300	18	17.8	35.8	3.19	3.35
1300–1700	18	16.8	34.8	3.17	3.33
1700–2200	18	16	34	3.16	3.32
2200–3200	18	15.5	33.5	3.14	3.30
3200–4400	18	15.2	33.2	3.13	3.28
4400 & above	18	14.8	32.8	3.12	3.27

Notes:
1. Costs based on surveys of representative jobs.
2. Percentages of costs have to be varied according to the individual contractor's experience and the type of project being estimated.
3. Supervision to be added as a direct job expense.
4. Above tables are contractors' estimated costs. No allowance has been made for return or profit.

FIG. 17-10. Printed lists showing sliding scales of costs are useful in establishing legitimate charges for small jobs, change orders, and "extras."

detailed figures. The estimated time and rate per hour for engineering and drafting must be provided. And so for any other expense, there must be detailed figures.

Very often buyers will not stop to study detailed figures. The fact that the contractor has them available will suffice. Customers are usually willing to pay for what they get but want to know what they are paying for.

After the contractor has justified the charges for his services, he still has to enlighten the buyer on the value of such services.

CHAPTER 18

Do You Know What It Costs to Train Apprentices?

THE COST OF TRAINING

The electrical contracting industry is constantly confronted with a challenge to supply mechanics vested with knowledge and skills equal to the demands of modern electrical construction. These demands become more exacting from year to year, and it requires much effort and astute planning on the part of electrical contractors to keep abreast of the times. To meet such demands, one of the first steps is the careful selection and proper training of men.

Carefully planned apprentice training has long been recognized as an indispensable institution in the electrical contracting industry. Since much has already been written on the subject, many phases will be omitted here.

Three sweeping questions will in a general way set forth the problems to be considered in this chapter:

1. What does it cost to train apprentices?
2. Who bears the burden?
3. Who benefits?

An analytical study of the accompanying chart "Estimated Cost of Apprentice Training—Electrical" will enable us to bring out many points which will later assist in answering the above questions. Before the reader can appreciate the figures used, however, he must understand what the term *apprentice training* is intended to embrace.

Scope of Training

As it will be used here, the term apprentice training means *proper training in all branches of electrical construction normally encountered by electrical contractors*. A present-day graduate of an apprentice training course, properly administered, has a better than average education and is a skilled mechanic. He has a good knowledge of elementary electricity and some knowledge of advanced electrical problems, is familiar with building construction practices, and is skilled in making electrical installations.

The figures used in the columns of the accompanying chart are based on the training being properly carried to completion in the 4-year training period. This training must be paid for, regardless of when the mechanic gets it. If it is not given during the regular training period, it will have to be paid for at a higher rate at some later time.

Apprentice training continues for 4 years. To provide for gradual increases in salary, each year is divided into four 13-week periods. There are 16, in all, of these 13-week periods, and the training cost is figured for each.

The term *estimated* is used in the heading of the chart. Such costs, as those for training mechanics, have to be estimated, as they involve items for which exact costs cannot be established. There is no means of accurately measuring the productivity of an individual working in a gang on varied forms of construction. Nor is it practicable to try to keep exact record of the time used by the regular mechanics in training new men. The cost to the job of having one of the crew periodically drop out a day for school is also a debatable amount. The more advanced the apprentice gets, the more costly this interruption of work becomes.

Division of Costs

For the purpose of study, three divisions of cost "estimated" were made, namely:

1. Value of services (Cols. 8 to 9)
2. Expense of training (field costs, Cols. 14 to 15)
3. Carrying costs (Col. 16)

These divisions are designed to cover the whole of training expense. The sum of the three costs is shown in Cols. 17 and 18. *The chart (Col. 18) shows that for the first 3 years the contractor's expenses exceed the dollar value of the services.* Column 17 indicates that for each period of the last year, the contractor gains.

ESTIMATED COST OF APPRENTICE TRAINING—ELECTRICAL

YEAR	THREE MONTHS PERIOD (13 WKS.)	RATE OF PAY PER HOUR $	RATE OF PAY PER WEEK (AV. 36 HRS. PLUS $1.00 FOR 4 HRS. SCHOOL)	SALARY FOR PERIOD (13 WKS.) A	VALUE OF SERVICES % OF JOURNEYMAN SEE NOTE 1.	VALUE OF SERVICES PER WEEK $	VALUE OF SERVICES FOR PERIOD (13 WKS.) B	NET VALUE OF SERVICES B−A GAIN +C	NET VALUE OF SERVICES B−A LOSS −C	EXP. TRAINING MECHANICS TIME % OF WEEK	EXP. TRAINING MECHANICS TIME COST PER WEEK	EXP. TRAINING MECHANICS TIME COST 13 WEEKS D	PRACTICE MATRL. & TOOLS E	FIELD COSTS CONTR'S. BALANCE C,D,&E COMBINED +F	FIELD COSTS CONTR'S. BALANCE −F	CARRYING COSTS IN EXCESS OF NORMAL SEE NOTE 2. G	TOTAL COST CONTR'S. BALANCE F & G COMBINED +H	TOTAL COST CONTR'S. BALANCE −H	F	G	H
COLS.	1	2	3	4	5	6	7	8	9	10	11	12	13	14	15	16	17	18			
					INDUSTRIAL, COMMERCIAL, & RESIDENTIAL WORK																
1	1ST	0.50	19	247.	20	16.	208.	—	39.	4	3.20	42.	1.	—	81.	87.	—	168.			
	2ND	0.80	29.80	387.	25	20.	260.	—	127.	4	3.20	42.	5.	—	174.	83.	—	257.			
	3RD	0.90	33.40	434.	30	24.	312.	—	122.	4	3.20	42.	10.	—	174.	78.	—	252.			
	4TH	0.90	33.40	434.	30	24.	312.	—	122.	4	3.20	42.	10.	—	174.	78.	—	252.			
2	5TH	1.00	37.	481.	35	28.	364.	—	117.	5	4.	52.	15.	—	184.	74.	—	258.			
	6TH	1.00	37.	481.	40	32.	416.	—	65.	5	4.	52.	15.	—	132.	69.	—	201.			
	7TH	1.15	42.40	551.	45	36.	468.	—	83.	6	4.80	62.	15.	—	160.	65.	—	225.			
	8TH	1.15	42.40	551.	50	40.	520.	—	31.	6	4.80	62.	15.	—	108.	60.	—	168.			
3	9TH	1.30	47.80	621.	55	44.	572.	—	49.	6	4.80	62.	10.	—	121.	54.	—	175.			
	10TH	1.30	47.80	621.	60	48.	624.	3.	—	6	4.80	62.	10.	—	69.	49.	—	118.			
	11TH	1.45	53.20	692.	65	52.	676.	—	16.	5	4.	52.	5.	—	73.	42.	—	115.			
	12TH	1.45	53.20	692.	70	56.	728.	36.	—	5	4.	52.	5.	—	21.	37.	—	58.			
4	13TH	1.60	58.60	762.	80	64.	832.	70.	—	2	1.60	21.	—	49.	—	23.	26.	—			
	14TH	1.60	58.60	762.	85	68.	884.	122.	—	2	1.60	21.	—	101.	—	18.	83.	—			
	15TH	1.75	64.	832.	90	72.	936.	104.	—	2	1.60	21.	—	83.	—	13.	70.	—			
	16TH	1.75	64.	832.	95	76.	988.	156.	—	—	—	—	—	156.	—	5.	151.	—			
TOTALS				9380.			9100.	491.	771.			687.	115.	387.	1471.	835.	330.	2247.			
LOSSES DUE TO APPRENTICES DROPPING OUT (APPROX. 5% IN THE FIRST TWO YEARS)																		89.			
																	330.	2336.			

NET COST OF TRAINING $2006. (2336.−330.) SEE COL. 17 (+H) & COL. 18 (−H)

NOTES:—
1.— $80. PER WEEK WAS USED FOR JOURNEYMAN RATE (40 HRS.)
2.— CARRYING COSTS, NOT INCL. IN TABLES, ARE EQUAL TO APPROX. 10% OF (JOURNEYMAN PAY PLUS TRAINING EXP.−LESS VALUE OF SERVICES)
  EXAMPLE—EST. FOR FIRST PERIOD= 10% OF [(1040+0+E)−B] = G10[(1040+42+0)−208] = 87.40 (87. USED)
3.— COSTS BASED APPRENTICE BEING WELL TRAINED IN ALL BRANCHES OF WORK DESIGNATED BY HEADING

ELECTRICAL CONTRACTORS ASS'N. OF CITY OF CHICAGO
R.A.

FIG. 18-1. It would be impractical to try to provide tables for all existing scales of wage rates. Here, $2 per hour ($80 per week) has been used because it provides a convenient multiplier when adjusting for other rates.

Value of Services

The index of the dollar value of an apprentice is the pay of a first-class journeyman electrician, which as used here is $80 per week. By estimating the productive labor as a percentage of that of a journeyman, the dollar value to the contractor is figured. Today, hourly rates may be higher, but the percentages and ratios used below still apply.

For the first period of training the productive labor is estimated to be 20% (see Col. 5) of that of a journeyman electrician. The value of the services per week is 20% of $80 or $16 (see Col. 6). For the 13-week period, the amount as shown in Col. 7 is $208. Column 4 shows that the contractor paid $247, or a premium of (247 − 208) $39.

Following the same sequence of figures for the thirteenth period (first quarter fourth year), we find that the contractor gains $70 (Col. 8). In the 4-year period the total salary paid is $9,380 while the value of the services is only $9,100. The contractor has invested $280 to cover this part of his expense of training.

Productive Efficiency

The percentage of productivity of apprentices is a much debated question. The diversity of opinions is largely due to lack of study of basic facts.

Column 5 shows a graduated scale starting at 20% for the first period and increasing to 95% for the final quarter of the fourth year. We know this does not depict actual conditions, but for estimating purposes it is the logical method to follow. It would be impossible to show actual percentages. An apprentice may follow a special line of work for a period and his productivity rises to 50%; then, upon being transferred to a new branch of work, it drops back to almost zero. His average for the two types of work would be 25%. *It is his average ability to perform all lines of work, that is represented by Col. 5.*

For actual installation work, most beginners are, for a time, a hindrance rather than a help. They can, however, aid in handling materials and tools, run errands, keep the storeroom in shape, and do other odd jobs. This useful work enables us to allow 20% for the first period.

One may learn how to perform a piece of work in a comparatively short time, but it may take weeks on the same type of labor before he attains any great degree of efficiency. House wiring is generally regarded as a simple line of electrical construction, yet any mechanic, trained or untrained, who is new to the work requires months before he can keep pace with a veteran residential wireman.

Weighing the complexity and number of all the new operations which the apprentice must study, practice, and master before he is a full-fledged electrician, one realizes that to complete the work in 4 years, he will have to be a bright and diligent workman. The progress must be rapid if the schedule of productivity shown in Col. 5 is maintained.

At the close of the war many young men were available whose education, training, and experience enabled them to both produce more and learn faster than the young apprentice starting from scratch. This was fortunate for the contractors as well as their customers, as such apprentices can complete the regular 4-year training in 3 years.

In order for apprentices to keep up with the percentage ratings shown in Col. 5, they require attention from well-trained mechanics. This supervision must be paid for by the contractor.

Expense of Training

Referring again to the chart, under the heading "Expense of Training," we find the two following items:

1. Mechanics' time (Cols. 10 to 12)
2. Practice materials and tools (Col. 13)

Each of these will be covered in its turn.

From the most menial task to the highly specialized services, every new piece of work undertaken by the apprentice requires attention from a trained workman. System and sometimes skill are required for unloading and handling material as well as for installing it. The beginner must be taught this.

There are days when the apprentice will go along with little attention, whereas other days much of the mechanic's time will be consumed explaining and demonstrating new operations. Often the instructor is obliged to stand by while the beginner tries his hand at the work. More costly than either (or both) instruction or lost time may be the reduction of the mechanic's own productivity due to working with an apprentice instead of another trained mechanic.

An average of 4% of the mechanic's time is charged to training over the 4 years. Suppose 2% were charged to reduction of the mechanic's productivity; that would leave 2% for training and instruction time. For an 8-hr day, 2% would be 9.6 min. There are times when it takes almost that long for the mechanic to figure out how to use the apprentice to the best advantage.

In addition to electrical work, there are all the other trades to be studied. The beginner must be taught how to read the plans of all other trades as well as the construction procedure and methods of such trades. He receives preliminary instructions in his school work, but such instructions must be augmented by on-the-job tutoring.

Practice Materials and Tools

The total amount charged to practice materials and tools is $115 (Col. 13). *This represents the amount in excess of normal costs for spoilage and use of tools.* The beginner must have materials to practice on and is bound to spoil some. Tools used by a novice are tied up longer than normal and subject to damage.

Since the total amount charged to tools and practice material is only $115, or a little over 50 cents (average) per week, it seems unnecessary to go into extended details on the subject.

Excess Carrying Costs

The cost of carrying an employee on the books is approximately the same regardless of rating. Normally, it costs in the neighborhood of 20% of the payroll to carry journeymen. To be very conservative, only half this amount, or 10% of journeymen's pay, was used as the carrying cost of an apprentice. Explanation of the method of determining excess carrying costs follows.

At $80 per week, the total journeyman pay for a 13-week period would be $1,040. Using the above percentage (10%), the normal cost of carrying the apprentice would be 10% of $1,040. The excess cost would be 10% of $1,040 *less the value of services*. This is shown in Col. 7 of the chart.

In addition, the contractor has a carrying cost on the "Expense of Training" item. Ten per cent of this amount is also used. The total excess cost (not included elsewhere) for the second period of training is figured as follows:

$$
\begin{aligned}
0.10 \text{ times } (\$1040 - \$260) &= \$78.00 \\
0.10 \text{ times } (42 + 5) &= 4.70 \\
\text{(See Cols. 12 and 13)} & \\
\text{Total excess carrying cost} &= \$82.70 \\
(\$83 \text{ used}) &
\end{aligned}
$$

Total Costs

The costs for each 13-week period are shown in Cols. 17 and 18. Column 17 shows the figures for the periods which netted a gain for the contractor, and Col. 18 shows the periods requiring an investment by the contractor.

The gain for the fourth year (Col. 17) is $330, and the total expenditure for the first 3 years, including losses due to apprentices dropping out, is $2,336 (Col. 18). The net investment by the contractor is therefore $2,336 − $330, or $2,006.

A net return of 10% on a $200,000 (cost) job would be required to pay the cost of training one apprentice. That is part of the cost of doing business, however, and must be accepted along with other costs.

WHO BEARS THE BURDEN

Apprentice training is a growing institution and must continue to grow if electrical construction is to keep abreast with the demands of the time. It is a means whereby the public can reap the greatest benefits from the man power expended.

In the previous paragraphs the cost of training electrical apprentices was discussed at length. Here, attention will be directed to the questions, "Who bears the burden?" and "Who benefits from the program?"

Two groups bear the burden of apprentice training: *the organized electrical contractors* who foster training programs and the *apprentices themselves*. The type of training which we have been discussing can be realized only by having carefully planned and rigidly administered schedules of instruction. It is a training intended to produce electricians adept in all lines of electrical-construction work.

The mechanic engaged on new building work must, in addition to knowing how to install his own work, be able to:

1. Read and interpret building construction plans
2. Visualize the completed project
3. Understand and anticipate the procedures and methods of other building trades
4. Time work to fit the job progress and provide proper coordination with other trades
5. Protect work during construction periods

The Apprentice's Investment

In order to master the school work and cope with other demands of a rigid course, the apprentice should have the equivalent of a high school education before entering the work.

The average age of the boys indentured, in normal times, is around eighteen to twenty years. In localities where rates for mechanics are comparable to those used in the tables previously discussed, chances are that an eighteen-year-old boy with the equivalent of a high school education could start at $40 or more a week in most occupations. With a $5 a week raise each year, he would be getting $55 for the fourth year. His average salary would be $47.50 per week. The total salary for the first 4 years would be $9,880 (208 × 47.50). The electrical apprentice earns $9,380 for the 4-year period. Using these modest figures for what he could *earn in other lines,* the amount he would have invested would be $500.

Any development of skills which provides for producing better work at lower costs is an economic good and, as such, benefits society in general. Apprentice training is not only an economic good; it has become an economic necessity.

The buying public always gains by having skilled mechanics to do its work. The contractor risks his investment, and the apprentice his future. The public risks nothing and always benefits.

One may say that the costs are passed on to the buyer. That is true to a certain extent, but in the end he gets value received and takes no gamble. In the past few years there has been an acute shortage of well-trained electricians, and the buying public has found this shortage very costly.

As stated in the first section, apprentice training is the lifeblood of the business and cannot be neglected. The apprentice elects his vocation and, in majority of cases, follows a natural aptitude for the work.

As contractors' appreciation of the benefits of apprentice training increases, their efforts to improve training programs also increase. In localities where contractors and the unions have formed joint training boards, both groups can benefit by having every graduate a certified man. A well-planned and carefully administered program can serve as a guarantee on anyone completing the 4-year course. The contractor, hiring such a man, will know he is securing someone suited to his work. The union, in turn, has its prestige raised by the high quality of such members.

For the apprentice engaged in electrical-construction work the signposts are optimistic. Many of the contractors, engineers, and estimators of today were the mechanics of yesterday.

Improved tools and construction equipment, together with improved and standardized materials, have greatly reduced the hardships and manual labor of electricians. In electrical-construction work we regard the properly trained electrician of today as a specialized technician. All things considered, the future of the present-day electrical apprentice looks bright.

CHAPTER 19

Can You Explain Your Tool Charges?

CUSTOMERS MAY DEMAND AN EXPLANATION OF SUCH CHARGES

Not too many years back, electrical contractors were giving little or no thought to tool costs. Tools were purchased, put on the books as overhead, and that was that. Prior to 1938, no extensive investigation of subject was made. In that year, a study of the cost of tools used in the electrical-construction industry was initiated in Chicago by Norman Pierce, Sr., president of Pierce Electric Company. As always, members of the Chicago Electrical Estimators Association were ready to assist in worthwhile research, and with their cooperation, the author prepared tables similar to those shown in Figs. 19-1 and 19-2.

The following year the author presented the results of the Chicago research in a paper entitled "Neglected Costs." Since that time the results of tool-cost studies have been widely promulgated in talks before electric contractor associations and by articles in the electrical-construction press. Now, the practice of charging for tools is becoming general among electrical contractors. Some figure approximately 3% of the labor cost; others include tool charges along with labor insurances, engineering, cartage, and other direct job costs.

Regardless of the method employed in billing, the contractor must be prepared to explain and defend his tool charges. You may say they are based on research. Suppose he asks you to substantiate your charges. You must be in a position to explain the basic factors which affect tool costs and understand how the figures are developed. Details of the procedure for studying tool costs will be covered in the succeeding paragraphs.

For example, your customer may pick an individual tool such as a power-driven threader and ask you to substantiate a charge of $23 per month for an item that has an initial cost of $300 and a nominal write-off of $2.50 per month for depreciation. The approximate life of such a tool is 10 years, or 120 months. Dividing $300 by 120 gives the $2.50 figure.

FACTORS AFFECTING TOOL COSTS

More than mere depreciation enters into the development of a tool-cost charge. Three specific factors must be dealt with when establishing tool charges. They are:

1. Time in actual use
2. Fixed costs
3. Individual job costs

Time in Use

One must not confuse the term *time in use* with *rated life*. Rated life is the period in which the tool may be available for economical use. Time in use is the time that the tool is actually in use or available for use on the job. The rated life may be 10 years, but the time in use may not be more than 3 or 4 years.

Let us study the possible time in use during a 10-year period for the power-driven thread cutter. Ten years serves well because it is the rated life of thread cutters, band saws, and many other heavy tools. Besides, the cycles in business are usually about 10 years' duration. Construction cycles generally follow a pattern of 7 good years and 3 bad ones.

Time in use for a power-driven thread cutter would then be estimated as follows:

7 good years at 60%.......	4.2 years
2 bad years at 20%........	0.6 years
1 year obsolescence........	0.0 years
Total time in use........	4.8 years
	or 57.6 months
	(Use 58 months)

Fixed Costs

The fixed charges on a tool consist of original purchase price, interest on investment, and insurance. When figuring the interest costs it is assumed that the depreciation is uniform over the 10-year period.

The cost of a power-driven thread cutter is estimated as follows:

1. Fixed costs:

Purchase price	$300.00
Interest on investment 5% for 10 years on $150 ($300 ÷ 2)	75.00
Insurance—usually carried by contractor— 0.5% for 10 years	15.00
Complete cleaning and painting and some overhauling (once every 2 years)—5 × $20	100.00
Total fixed cost for 10 years	$490.00
Time in use—58 months (based on normal construction cycle)	
Fixed costs for 58 months of actual use	$490.00
Fixed cost per month ($490 ÷ 58)	$8.45

2. Individual job expense—prorata cost per month while in use, based on 3 months on the particular job:

Cartage (out and back at $6 each) ($12 ÷ 3)	$ 4.00
Handling labor (out and back at $4.50 each) ($9 ÷ 3)	3.00
Repairs while on job—average $4 per month	4.00
Total prorata job expense per month	$11.00

   Summary of costs per month:

a. Fixed costs		$ 8.45
b. Individual job expense		11.00
General overhead plus insurance on Labor	$19.45	
(shipping and maintenance)—20%		3.89
Total cost per month		$23.35
Use $23 per month		

Hence, the total amount of interest is equal to the normal rate for 10 years on half the original purchase price (see above estimate).

The insurance is usually carried by the contractor. Nevertheless, it is an expense that must be included.

Individual Job Expense

Charged to the individual job are the expense of cartage and handling (transportation from shop to job and back) and costs of repairs. Repairs must be prorated. A tool may go through one or more jobs without repairs. Then, on another job, tool-repair expense may be substantial.

TOOL LIFE VARIES

The life of a given tool will vary considerably with the type of work. This is particularly true of ladders, ropes, fish tapes, and other tools subject to varying loads and abuses. A ladder used for trim and light conduit work will naturally stand up much longer than one used for heavy conduit or equipment. Ropes, not used properly, can have a very short life.

Some contractors erroneously think that tools have a long life because they lie around the shop for a long time. Idle tools in the shop and on the job may result from mechanics' reluctance to use them. No contractor should discount estimated tool-cost figures, such as those illustrated, until he can prove some error. There will be times when the estimated costs must be varied. Light intermittent duty may justify a reduction. For projects with a great deal of overtime, the costs must be increased.

1956 TOOL COSTS

For a Crew of 50

Figure 19-1 gives a list of tools for a job requiring a crew of 50 electricians. The list is compiled with the idea of getting a cost that normally would be experienced. It is by no means a listing of all the tools in common use on large projects. In addition to a long list of small tools that might be added, there are large shop tools, portable compressors, surveyor's levels, fork-lift trucks and numerous other costly items. To include all the tools that might be found in use on various jobs would produce an estimated cost in excess of common experience.

The estimated monthly cost of tools for a 50-man crew is indicated in Fig. 19-1 as $926. With a labor rate of $3 per hour and a 40-hr week, the cost would represent approximately 3.6% of payroll. For a $3.50 per hour rate, the monthly tool cost would approximate 3% of payroll. Tool cost also will vary with the type and size of project and the adequacy of tools.

Since 1945, the cost of labor has increased faster than the cost of tools. However, the cost of tools for a 50-man crew is about the same when expressed as a percentage of payroll. For smaller projects, the percentage of cost has increased, since the number of expensive tools on smaller jobs has increased. Also, there has been some increase in the use of moderate-priced tools. Greatest increase in tool usage appears to be in the categories of rolling scaffolds, wagon trucks, and lift platforms. A study of tool costs for a 5-man crew will show the effects of job size.

For a Crew of Five

The estimated tool cost per month for an industrial project requiring a crew of five electricians is shown in Fig. 19-2. The monthly cost

Item	Remarks	No. in use	Purchase price each	Total cost the lot	Depreciation and int. on invest. per mo. each - Per cent	Depreciation and int. on invest. per mo. each - Dollars	Estimated storage and repairs per month each	Total cost per month - Each	Total cost per month - The lot
Electric drills	¼ in.	4	$ 45.00	$ 180.00	5.5	$ 2.48	$0.50	$ 2.98	$ 11.92
Electric drills	½ in.	2	70.00	140.00	5.5	3.85	0.60	4.45	8.90
Stepladders	8 ft	6	15.00	90.00	20	3.00	....	3.00	18.00
Stepladders	12 ft	8	24.00	192.00	25	6.00	0.50	6.50	52.00
Stepladders	14 ft	5	35.00	175.00	30	10.50	0.50	11.00	55.00
Extension ladders	32 ft	1	35.00	35.00	10	3.50	0.25	3.75	3.75
Pipe benches	Small No. 53	5	23.00	115.00	5	1.15	....	1.15	5.75
Pipe benches	Large No. 54	5	30.00	150.00	6	1.80	....	1.80	9.00
Pipe vise	½ to 2 in.	5	11.00	55.00	4	0.44	0.25	0.69	3.45
Pipe vise	2 to 3½ in.	5	19.00	95.00	4	0.76	0.50	1.26	6.30
Stocks	½ to ¾ to 1 in.	6	17.00	102.00	10	1.70	....	1.70	10.20
Stocks	1¼ to 2 in.	6	25.00	150.00	10	2.50	....	2.50	15.00
Stocks	2-in. ratchet	3	38.00	114.00	10	3.80	0.25	4.05	12.15
Stocks	2½ to 4 in.	3	115.00	345.00	5	5.75	2.00	7.75	23.25
Thread cutter	Power drive	2	300.00	600.00	5	15.00	8.00	23.00	46.00
Pipe dies	½ and ¾ in.	10	3.60	36.00	60	2.16	....	2.16	21.60
Pipe dies	1 and 1¼ in.	6	5.50	33.00	60	3.30	....	3.30	19.80
Pipe dies	2	4	6.00	24.00	40	2.40	....	2.40	9.60
Pipe dies	2½ in.	3	10.00	30.00	25	2.50	....	2.50	7.50
Pipe dies	3 in.	3	10.00	30.00	25	2.50	....	2.50	7.50
Pipe dies	4 in.	1	12.00	12.00	20	2.40	....	2.40	2.40
Hickies	½ to 1 in.	15	5.00	75.00	10	0.50	....	0.50	7.50
Power saw	Large band	1	710.00	710.00	5	35.50	8.00	43.00	43.00
Pipe benders	Small	2	202.00	404.00	5	10.10	2.00	12.10	24.20
Pipe benders	Large	1	600.00	600.00	5	30.00	4.00	34.00	34.00
Tool boxes	Wood	4	35.00	140.00	10	3.50	0.50	4.00	16.00
Tool boxes	Steel	8	35.00	280.00	5	1.75	....	1.75	14.00
Reel dollies		3	35.00	105.00	5	1.75	....	1.75	5.25
Winches	Hand driven	2	80.00	160.00	5	4.00	1.00	5.00	10.00
Winches	Power driven	2	180.00	360.00	5	9.00	4.00	13.00	26.00
Jacks	Stone	2	150.00	300.00	4	6.00	....	6.00	12.00
Jacks	Reel	12	35.00	420.00	4	1.40	....	1.40	16.80
Fish tape	⅛ in.—100 ft	6	2.75	16.50	50	1.37	....	1.37	8.22
Fish tape	3/16 in.—100 ft	4	3.50	14.00	50	1.75	....	1.75	7.00
Fish tape	¼ in.—200 ft	2	8.50	17.00	25	2.12	....	2.12	2.24
Blocks	Snatch and misc.	6	7.00	42.00	5	0.35	....	0.35	2.10
Rope—hemp	½ in.—200 ft	2	8.00	16.00	30	2.40	....	2.40	4.80
Rope—hemp	¾ in.—200 ft	1	15.50	15.50	30	4.65	....	4.65	4.65
Rope—wire	¼ in.—150 ft	2	25.00	50.00	10	2.50	....	2.50	5.00
Rope—wire	⅜ in.—200 ft	2	40.00	80.00	10	4.00	....	4.00	8.00
Chain hoist	1—5-ton 10-ft lift	1	270.00	270.00	3	8.10	....	8.10	8.10
Wagon trucks		3	50.00	150.00	8	4.00	1.00	5.00	15.00
Whitney punches and dies	Misc. sets	4	65.00	260.00	3	1.95	2.00	3.95	15.80
Extension cords—H.D.	50 ft—complete	16	8.00	128.00	15	1.20	0.50	1.70	27.20
Motor generator set		1	195.00	195.00	5	9.75	3.00	12.75	12.75
Gas furnace	Plumbers	2	35.00	70.00	4	1.40	1.00	2.40	4.80
Gas tanks and burners	Compressed gas	4	50.00	200.00	4	2.00	....	2.00	8.00
K.O. punches		4	10.00	40.00	20	2.00	....	2.00	8.00
Pipe wrenches	18 in.	10	5.00	50.00	10	0.50	....	0.50	5.00
Pipe wrenches	24 in.	8	8.75	70.00	10	0.88	....	0.88	7.04
Chain tongs	Misc.	3	Avg 18.00	54.00	10	1.80	....	1.80	5.40
Cable pullers	Comealong	6	20.00	120.00	50	6.00	....	6.00	36.00
Scaffolding	Rolling and misc.	..	300.00	300.00	10	30.00	4.00	34.00	34.00
Star drills		20	1.00	20.00	20	20.00/c	0.20	0.20	4.00
Reamers	Misc. ratchet	6	15.00	90.00	10	1.50	0.50	2.00	12.00
Files	Misc.	50	0.70	35.00	50	0.35	....	0.35	17.50
Twist drills	Misc.	80	0.50	40.00	25	0.13	....	0.13	10.40
Taps	Misc.	50	0.60	30.00	60	0.36	....	0.36	18.00
Oilers		20	0.50	10.00	30	0.15	....	0.15	3.00
Hammers and sledges	Misc.	6	4.00	24.00	10	0.40	0.50	0.90	5.40
Bull points	Misc.	6	2.00	12.00	10	0.20	....	0.20	1.20
Locks and chains		20	4.00	80.00	8	0.32	....	0.32	6.40
Hacksaw blades		300	0.10	30.00	100	0.10	....	0.10	30.00
Shop supplies and misc.	Waist, oil, etc.	Lot	20.00	20.00	100	20.00	....	20.00	20.00
Shop tools	Grinders, etc.	Lot	100.00	100.00	10	10.00	1.00	11.00	11.00
Totals		..	....	$8,906.00	.....	.....	....	.....	$925.42

NOTE: Storage includes prorata shipping and handling costs. Based on an hourly rate of $3.50, the total of $926 would represent approximately 3 per cent of the payroll.

FIG. 19-1. Estimated cost of tools for a project requiring 50 electricians showing a cost breakdown per tool item with monthly cost analyses. Note depreciation, interest, storage, and repair factors.

ESTIMATED COST PER MONTH FOR TOOLS TO EQUIP AN INDUSTRIAL
JOB REQUIRING A CREW OF FIVE ELECTRICIANS

Item	No. in use	Purchase price Each	Purchase price Lot	Job cost per month Each	Job cost per month The lot
Electric drills—¼ in.	1	$ 45.00	$ 45.00	$ 2.95	$ 2.95
Electric drills—½ in.	1	70.00	70.00	4.45	4.45
Stepladders—8 ft.	1	15.00	15.00	3.00	3.00
Stepladders—12 ft.	2	24.00	48.00	6.50	13.00
Stepladders—14 ft.	1	35.00	35.00	11.00	11.00
Pipe benches—small	2	23.00	46.00	1.15	2.30
Pipe benches—large	1	30.00	30.00	1.80	1.80
Pipe vise—½ to 2 in.	2	11.00	22.00	.70	1.40
Pipe vise—2 to 3½ in.	1	19.00	19.00	1.25	1.25
Stocks—½ to 1 in.	2	17.00	34.00	1.70	3.40
Stocks—1¼ to 1½ in.	1	25.00	25.00	2.50	2.50
Stocks—2-in. ratchet	1	38.00	38.00	4.10	4.10
Stocks—2½ to 4 in.	1	115.00	115.00	7.75	7.75
Pipe dies—½ to ¾ in.	4	3.60	14.40	2.15	8.60
Pipe dies—1 to 1¼ in.	2	5.50	11.00	3.30	6.60
Pipe dies—2 in.	1	6.00	6.00	2.40	2.40
Pipe dies—2½ in.	1	10.00	10.00	2.50	2.50
Pipe dies—3 in.	1	10.00	10.00	2.50	2.50
Pipe dies—4 in.	1	12.00	12.00	2.40	2.40
Hickies—miscellaneous ½ to 1 in.	5	5.00	25.00	0.50	2.50
Pipe benders—small	1	200.00	200.00	12.00	12.00
Toolboxes—large wood	1	50.00	50.00	5.00	5.00
Toolboxes—steel	2	35.00	70.00	1.75	3.50
Rope—hemp ½ in. 100 ft.	1	4.00	4.00	2.00	2.00
Rope—hemp ¾ in. 100 ft.	1	7.50	7.50	5.00	5.00
Scaffolding—rolling	1	250.00	250.00	20.00	20.00
Wagon trucks	2	50.00	100.00	5.00	10.00
Chain hoist—5 ton, 10 ft lift	1	270.00	270.00	6.00	6.00
Extension cords—50 ft heavy duty	2	8.00	16.00	1.50	3.00
Gas furnace—plumbers	1	35.00	35.00	2.50	2.50
Pipe wrenches—18 in.	2	5.00	10.00	0.50	1.00
Pipe wrenches—24 in.	2	8.75	17.50	0.90	1.80
Chain tongs	2	18.00	36.00	1.75	3.50
Reamers—miscellaneous ratchet	3	15.00	45.00	2.00	6.00
Oilers	4	0.50	2.00	0.15	0.60
Knock out punches	1	10.00	10.00	2.00	2.00
Locks and chains	4	4.00	16.00	0.30	1.20
Hacksaw blades	30	0.10	3.00	0.10	3.00
Files—miscellaneous	5	0.70	3.50	0.35	1.75
Miscellaneous—hammers, drills, and miscellaneous shop supplies			20.00 5.00		2.00 5.00
Total			$1,800.90		$183.25

FIG. 19-2.

of $183 represents approximately 6% of a $3.50 per hour payroll. In 1945, the estimated cost of tools to equip such a job was about 5.5% of a $1.70 per hour payroll.

As time goes on, the gap between the cost of tools for large and small jobs will be greater (on a percentage of payroll basis). Contractors find that expensive man-power saving tools are economical on small as well as large projects. Also, mechanics become accustomed to working with the better tools and expect to find them on the job.

As illustrated in Figs. 19-1 and 19-2 the cost of tools per man-day is much greater for a 5-man crew than it is for a 50-man crew. Studies show that the cost levels off as the size of the crew approaches 50 men. Above the 50-man level, there is little change. There are several reasons for this. The principal one is that a 50-man crew proves to be an economical unit.

BILLING FOR TOOLS

In competitive bidding, tools are included in the estimate with other direct job costs. On cost-plus work, they may be billed:

1. As a percentage of payroll
2. On a rental basis
3. As a material item (sold directly to the job)

The simplest method of billing is that of charging a percentage of payroll. Some contracts call for tools to be billed on a rental basis. In that case, so much per month is billed for each of the depreciated tools. Consumed tools, such as drills, taps, hacksaw blades, and similar items, are charged directly to the job the same as material.

On many large cost-plus projects, the tools are purchased new and charged directly to the job. At the completion of the contract, these tools are turned over to the owner or the contractor buys them at a determined salvage price.

We have been studying industrial projects. Cost of tools generally runs higher for office and other commercial building electrical-construction work. Tools are tied up longer for the same man-hours on this type of poject.

Each contractor must study his own costs and figure tools accordingly. He *must not overlook* any of the many expense items involved in the supply and maintenance of tools. They represent a substantial item of job cost and must not be treated as an incidental item of overhead.

CHAPTER 20

Do You Understand Your Labor Curves?

Graphic representation has become a universal medium of expression in the business world. Engineers show motor and machine performance data in the form of plotted curves. Management relies on this method to plot its production and sales progress. Over the years, electrical contractors have studied and developed curves of various types to guide them in the efficient operation of their business.

One of the most effective to the contractor is a set of man-power demand curves. Such a management tool will show him the way to selection of the better jobs, reflect the effect of good and poor management on individual projects, and often explain profits and losses.

To get the full significance of the curves, generally it is necessary to plot the MPD (man-power demands) of several projects which overlap. Frequently the requirements of one job will affect the shape of a curve on others. Valleys and humps may be caused by shifting men back and forth to satisfy the temporary demands of other jobs. However, no attempt will be made to cover over-all volume at this time. Attention will be confined to factors which influence individual projects.

POOR MANAGEMENT

The curve in Fig. 20-1 shows the effect of poor management. The dashed line represents estimated labor demands, the solid line actual man power used on a feed-mill project. Estimated time was based on good management and close supervision of the work—a factor not realized on this particular job.

When the architect's representative called for men, the contractor was too busy to give the request his personal attention and decided to let the mechanics get the work started. He would visit the job presently—a resolve that was constantly postponed. In the interim, everyone except the contractor took part in running the work.

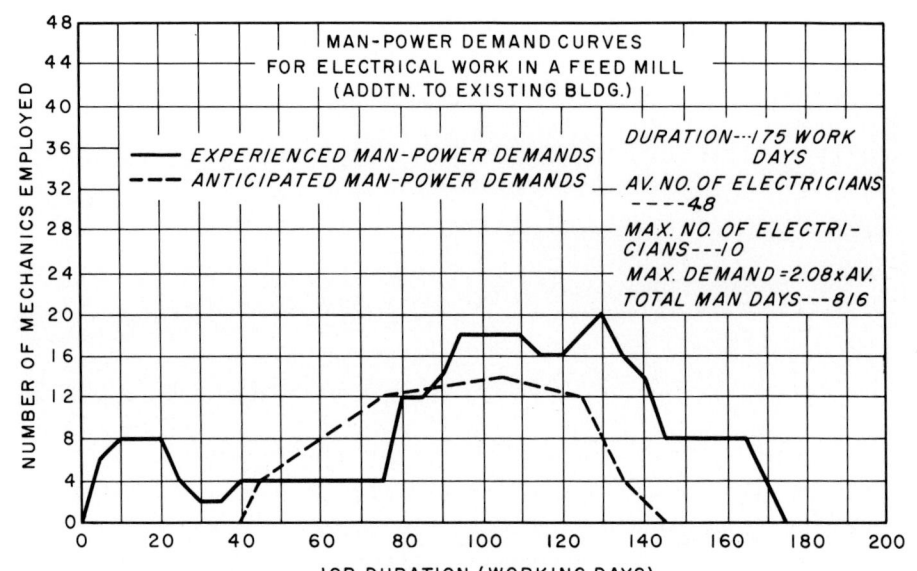

Fig. 20-1. Job neglect is indicated by these man-power curves on a feed-mill electrical project. Note the striking difference between estimated job duration and actual man-hours expended, also the excessive peaks.

Actually, the architect's representative insisted on having men on the job before they were needed. He also fussed and fumed whenever the contractor's foreman wanted to let any men go. During this time, other trades had electricians at their beck and call. As long as the electrical contractor gave the project no attention, his foreman was willing to let it drift along overmanned.

A "post-mortem" analysis revealed that men were on the job two weeks before they were actually needed and most of the time the number of mechanics was double that required. The maximum number of men used was ten, whereas six would have been sufficient.

The hump at the beginning of the man-power demand curve in Fig. 20-1 shows what happens when one tries to "push" a job. An electrical-construction project can go just so fast, and any attempt to push it takes a toll in labor. It is better to have too few than too many men on a job.

Do You Understand Your Labor Curves? 211

The electrical contractor is always confronted with the problem of striking a balance between good service and economic business practice. He must try to keep architects, engineers, and buyers satisfied, but he cannot afford to cater to any whims which are costly from the standpoint of labor.

Anyone with authority on the job is likely to try to tell the contractor when and how many men are needed. Unless the contractor

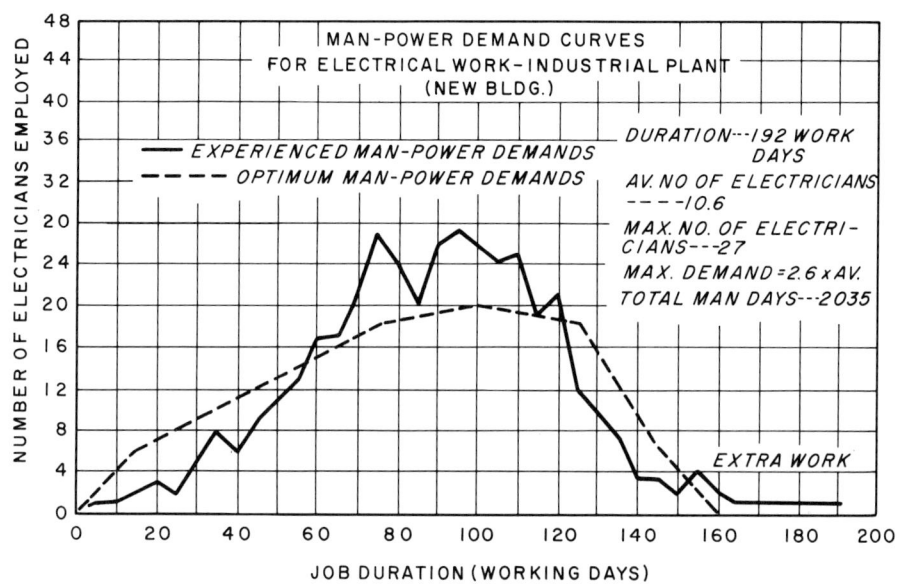

FIG. 20-2. Severe peak demands show up on this man-power demand curve for an industrial-plant project. Peak demands should approximate 1.75 times average demands.

visits the job site often enough, he is in no position to refute claims of others gracefully. There are times when a contractor must take a firm stand. He must run his own work.

Much has already been written about "optimum duration" (see *Electrical Estimating,* McGraw-Hill Book Company, Inc.); hence this discussion will be limited to a few comparisons. Data on an industrial project are shown in Fig. 20-2. Actual man-power demands are represented by the solid-line curve. The optimum-condition curve is plotted with a dashed line. In general, the actual demand curve compares favorably with the optimum curve. Duration periods coincide. Aside from the heavy peak demand, other variations are not too pronounced.

Because the man-power build-up shown by the solid line in Fig. 20-2 was gradual, it was not serious. The peak demand, however, was severe and indicated a serious condition because the job had either to draw heavily on other projects for men or to recruit mechanics from the open market. The seriousness of the latter alternative depended upon the prevailing condition of the labor market.

The maximum demand for the project in Fig. 20-2 was 2.6 times the average demand. Actually, a healthy state exists on a job when maximum man-power demand approximates 1.75 times the average demand.

A dip in the curve, between the seventeenth and eighteenth weeks (75 and 85 days) was by no means good. It was the result of unexpected delays in material deliveries. The dissolution period (120 to 150 days) was a trifle abrupt, but not serious. If men are laid off too fast, other projects cannot absorb them.

On the whole, the project in Fig. 20-3 reflects good management and shows the opportunities of industrial work. Such projects generally provide enough work that can be carried on independently of other trades to enable the electrical contractor to maintain an even flow of labor.

COORDINATION

Of course, there are types of industrial contracts which require much of the electrical work to be closely coordinated with that of other trades. Oil refineries are of this type.

A man-power demand curve for an oil refinery electrical project is shown in Fig. 20-3. Again, the curve does not depart too radically from the optimum curve. Had it not been for layoffs due to procurement failures (materials not on job when needed) and holiday absentees, the hump undoubtedly would have been appreciably reduced.

VOLUME AND OTHER FACTORS

Irregularities on one project may be caused by trying to satisfy the irregular demands of other jobs. Hence, it is frequently necessary to study, simultaneously, the man-power demand curves of several completed contracts. The greater the volume of work, the less likely that individual contracts will suffer from the fluctuating demands of others. To avoid the serious effects of varying labor demands, the electrical contractor must operate within his limitations.

The type of project, management of other trades, weather conditions, and numerous other factors may be reflected in the labor curves. Office buildings, sewage-disposal plants, warehouses, and most other

Do You Understand Your Labor Curves?

types of construction have their individual man-power requirements. Contractors are aware of this but may not appreciate the details of such requirements until they study the labor curves. The electrical contractor knows that management of the general contractor and other trades will affect his own labor requirements. Again, labor curves will indicate the seriousness of such influences.

In times such as these, when new architectural firms are springing up and established firms are overexpanding, it would be wise for the

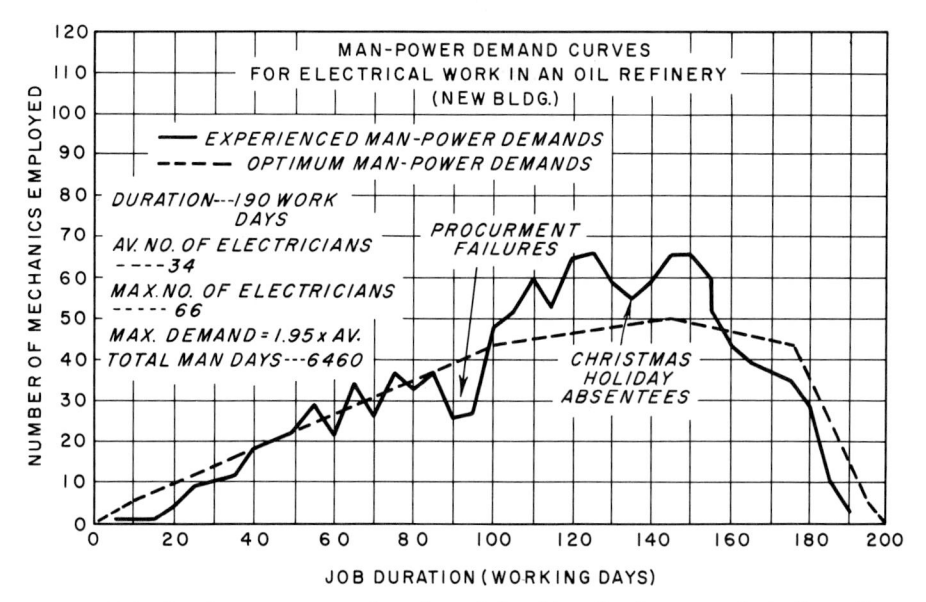

FIG. 20-3. Good management is reflected in this oil-refinery project. Excepting dips and peaks due to procurement failures and holiday absentees, labor curve closely approximated estimated optimum demand curve.

electrical contractor carefully to appraise the ability of any firm to which he plans to submit an estimate. Another factor is the speed with which plans are prepared. Drawings may be rushed to completion and contracts let. Once the work is started, mistakes crop up and progress is invariably retarded. Regardless of the type of formal contract or agreement, contractors find such projects costly. Foremen and key men are tied up with a consequent increase in overhead. Such influences show up on man-power demand curves as low, flat sections.

An existing project, with which the author is familiar, is a good example. It is dragging along owing to a change in design of steel columns. The work started in a big way and all trades had their crews fully organized. Suddenly it was discovered that the column design had to be changed, and the job progress virtually came to a halt.

When the man-power demand curves for this project are plotted, they will show a rapid build-up, a short hump, and a sudden drop followed by a long, flat section.

In good business practice, it is imperative that a contractor plot labor demand curves for some of his work. Not all jobs need such attention. However, in a widely varied selection of contracts, there are always many projects which should receive detailed analysis. There is no substitute for graphical representation when seeking a means of portraying certain specific details.

Plotting performance curves is nothing new. It has been considered an important part of good business practice for many years. Those electrical contractors who have been following this practice have found their efforts richly rewarded.

CHAPTER 21

Watch Your Trim Labor

Why do jobs lose money on the *trim?* That question has been asked repeatedly over the years. At a meeting of electrical contractors in Houston in 1949, the question was advanced again. It seemed to be the general consensus that jobs often progressed favorably through the roughing (installing of conduit, boxes, and associated materials) stages and then began to slump off.

The term *trim* is used loosely in the electrical contracting business, and the work it embraces is not generally defined. For our purpose, let us say that it includes all work that comes after roughing-in and pulling of feeders have been completed. Specifically, it includes the installation of branch-circuit wire, panels, wall switches, convenience outlets, signal equipment, motors, and other operating equipment plus testing of same.

On single-family and small multiple-unit residential buildings, wire pulling of all kinds is classed as trim work. Cable pulling is commonly considered trim work by contractors, but for our discussion we shall omit pulling of feeder and service cables.

The subject of "trim labor" is often queried but seldom studied in detail. Delving into it can prove enlightening.

A study of several projects reported as having had excessive costs for the trim labor revealed that very often one or more of the following conditions existed:

1. Analysis of job progress had been faulty
2. The job had been poorly managed

3. Labor estimated for the trim work was not adequate
4. The mechanics had "let down" on the work

Let us study each condition separately.

FAULTY ANALYSIS

It is common for electrical contractors to overestimate the portion of a job that is completed. This is particularly true of roughing-in labor. The job is ready for cable pulling and setting panels. It is assumed, therefore, that the roughing-in labor is all completed. In reality there may be numerous costly odds and ends that have to be picked up. Besides, there is the prorata share of cleanup and shipping-out labor that should be charged to roughing-in. Cleaning cabinets and boxes, scattered piecing out of conduits, removing conduit obstructions, installing box covers, and shipping out are all items of cost which are frequently and incorrectly charged against trim labor. Such accounting reflects cost all out of proportion to true values. Some figures will emphasize this.

On a hospital job, with which I am familiar, 70% of the labor was consumed for roughing-in and pulling of feeders; the remaining 30% was trim labor. On the basis of 100 hr, the corresponding divisions were 70 and 30 hr. If the roughing-in and cable work had been reported complete when it was only 85% done, there would have been 15% of same charged against the trim. This would be 15% of 70 hr or 10.5 hr. This is over 33% of the total time required for trim labor. Naturally the report on trim labor would have looked very bad if it had been burdened with an excess charge of over 30%.

Contractors might become cognizant, to some degree, of their mistaken appraisals if they followed jobs closely to completion. However, they are prone to slight the work as soon as it has passed the roughing-in stage. Both false appraisals and job neglect are problems for management.

Management plays an important part at every stage of a construction project. However, the discussion will be confined to the trim portion. Work must be followed through to completion. The hours wasted at the finish of a job are just as valuable as those lost at the beginning. Every dollar added to the cost of the job makes two dollars difference in the showing on the books. It adds one to cost and takes one away from profit.

One of management's most costly mistakes is the failure to reduce crews before the jobs begin to drag. A project must be followed closely

Watch Your Trim Labor 217

so that as soon as it passes the peak, removal of mechanics will be in step with the decline of work. On overmanned jobs the foreman cannot use his mechanics to the best advantage and they, in turn, slacken their pace. A surplus of men on a project often gives rise to false and detrimental rumors such as: "This is a cost-plus job"; "We are way ahead on labor"; or "Men are scarce and the boss doesn't want to let us go."

Fig. 21-1. Follow job progress closely by making regular visits to the project and consulting with the construction superintendent. Adjust size of crew to fit the work at hand.

There is a tendency to have a dislike for pickup work, and it is put off as long as possible. Again, there are foremen who think that minor unfinished tasks will serve as good fill-ins for "rainy days" or slack days. Those days never come, and as the main contract nears completion, the unfinished tasks pile up.

One Chicago contractor, in making his regular inspection of job progress, notes any unfinished odds and ends of work. His notes are turned over to the foreman with instructions that the work be taken care of at once. He contends that this program not only aids in keeping an accurate accounting of the job progress but actually saves labor hours.

During the progress of the main contract, men assigned to pickup work want to hurry up and get rid of it. At the end of the contract, the same work would be dragged out.

There are times when contractors deliberately keep men on a job when not needed to hold them for future work. In such cases the current job should be honestly noted as carrying excess labor and not be considered a dragging contract.

Fig. 21-2. Management should plan a job carefully before construction begins, set up a timetable for labor based on known and anticipated job conditions, then follow it as closely as is practicable.

Another management fault lies in having wrong men on the job for the trim or finishing work. To overcome this difficulty, some contractors have their mechanics divided into two groups: one group for roughing-in and the other for trim work. Shifting of crews is limited to large jobs.

The labor market and the volume of work put certain limitations on the contractor's flexibility in selecting men. This must be taken into account when the work is being estimated. The contractor has no right to complain about job progress just because he failed to estimate the work according to prevailing conditions.

There are two reasons why labor units used for trim work are frequently too low. First, they are of the stop-watch type. Second, upward adjustments of units to meet rising labor costs are centered on the roughing materials. The so-called "stop-watch" labor unit obtained by spot checking job labor is confined to concentrated portions of the work, and does not give a true picture of over-all costs. No allowance is made for scattered portions of the work, and preparation and pickup time are not accounted for.

In the event of a poor labor market, contractors try to adjust their labor units accordingly. Too often attention is concentrated on conduit, boxes, and large cables. Trim materials are overlooked. A similar faulty adjustment occurs when preparations are being made to figure complicated projects. The slow progress of sewage-disposal plants, railroad stations, and similar complicated jobs is not limited to the roughing-in work; it prevails right up to the very end. Contractors are prone to overlook its impact on trim labor.

Estimators ignore the effect that wire size will have on the labor for installing wall switches, receptacles, and similar devices. They use units based on No. 14 wire to figure work where the minimum size of wire is No. 12 or larger.

MECHANICS' "LETDOWN"

Fatigue, apathy, and fear of being out of work are among the principal reasons for mechanics "letting down" on the job. Fatigue may be from outside causes, but we shall limit ourselves to the job causes. Too much pressure during the roughing-in stages of the job or long hours may tire men to a point where they cannot do their best. The trim portion of the work, being last, suffers most from the effects of overworked mechanics.

In electrical construction, as in all other lines of work, there are men who avoid doing more than they have to. The nature of such mechanics varies from apathetic to downright ornery. Many mechanics have a feeling, and some rightfully so, that their bosses are getting all they can out of them.

It isn't natural for a man to work himself out of a job. When men are being laid off instead of being transferred to other projects, those remaining are inclined to slow down. In the absence of close supervision, the impression may evolve that management wants the mechanics to mark time pending the opening up of another job.

Unless the labor market is so bad that the poorer mechanics must be used, management can forestall this situation.

PSYCHOLOGICAL CONDITIONS

Contractors are always anxious to see jobs completed at an early date. Because of this they are frequently prone to label as "dragging" a project which, in reality, is progressing normally. Also, there is an abundance of "wishful thinking" in the electrical-construction fraternity. A venture moves along smoothly in the early stages, and the contractor immediately decides that he is going to make a "killing." When things settle down and he realizes that the profit will not approach his expectations, he complains that the work dragged along in the trim stages.

Jobs naturally slow down in the finishing stages. It is not a case of a letdown; it is the effect of working against odds. Allowances must be made for this in the estimate. If management is functioning properly, there is no reason for jobs to lose money on the trim work.

CHAPTER 22 *

The Wide Bid Spread

A marked variation in bids for electrical work does not necessarily mean that bidding is "wild" or careless. I have studied many cases where a wide spread existed between submitted prices and found very few instances that could justifiably be termed wild bidding. Often, it is a case of the low bidder being just a few jumps ahead of his competitors.

There are four common causes for a wide variation in bids. These include the following:

1. Condition of a contractor's business
2. Faulty markups
3. Resourcefulness of individual contractors
4. Rushed estimating

CONDITION OF BUSINESS

Bidding is naturally affected by the condition of the individual contractor's business. The man who has jobs nearing completion and nothing ahead will, out of necessity, figure more closely than his competitors who have plenty of work on their books. Variations in bids from contractors representing these two positions may seem startling on the surface. However, when carefully studied, they are found to reflect normal procedures.

To illustrate, let us set up a hypothetical case. *A* represents a con-

* Reprinted from an article in *Electrical Construction and Maintenance,* October, 1956.

tractor in need of business, and *B* a contractor with plenty of work. They are bidding on a job that would normally have an estimated base cost of $10,000 with a 60/40 M/L ratio (base cost 60% material and 40% labor). The base costs would be $6,000 for material and $4,000 for labor.

A soon will have his better mechanics available, and with ample engineering and supervision he can cut the normal labor down by at least 5%. He has $3,800 for the estimated cost of labor instead of the normal $4,000.

Contractor *A* also is willing to use some of the reserve that has been set aside for dull times. Instead of the normal overhead markups of 10 and 35% for material and labor, respectively, he uses 5 and 25%. For return he adds 5% to material and 10% to labor.

Contractor *B*, on the other hand, is not anxious to get the work but feels obliged to furnish a price. He tells his estimator not to figure too closely because their labor situation is not good and additional work would interfere with existing jobs.

After hurrying through the job, *B*'s estimator comes up with estimated costs of $6,300 for material and $4,200 for labor. To make sure the labor is ample, *B* adds another $200. Markups of 35 and 10% are applied to labor and 10 and 10% to material.

The accompanying Figs. 22-1 and 22-2 show the estimates as prepared by contractors *A* and *B*, respectively. Figure 22-3 gives the amounts that would normally be used in estimating the job.

Final estimated costs, as shown are:

Contractor *A*	$12,600
Contractor *B*	15,125
Normal estimate	13,710

Contractor *A*'s bid is $2,525, or 16.7% less than contractor *B*'s. There is a substantial gap between the two bids, yet from the standpoint of good business practices, there is nothing radically wrong with either.

A's bid is $1,110 below the normal estimate. However, there was a $200 saving in the base cost of labor. To this must be added the markups of 25 and 10%, plus the allowance for insurances and job expense of $40 plus 5%. The total savings on labor and labor services adds up to $317.

Deducting $317 from $1,110 leaves a balance of $793, which represents the amount that *A*'s bid is actually lower than it normally would be. In percentage the bid is approximately 5.8% lower than normal. This amount would still further be reduced if the normal allowance of 0.75% for reserve were deducted.

PRICING SHEET

CHI. ELECT. EST. ASSOC. FORM 4

DATE **Nov. 1** 19___

JOB OR BLDG. **DILLARD PRODUCTS**   LOCATION **1520 HIGH ST.**
BID TO **Do.**   ADDRESS
ARCHT. OR ENG'R.   ADDRESS
PLANS MARKED   SCALE   SPEC. No.   EST. No.
EST. BY **L.K.**   PRICED BY   EXTENSIONS BY   CHECKED BY   S.O.No.   SHEET No. **125**

MATERIAL	QUANTITY	MATERIAL UNIT	EXTENSION	LABOR UNIT	EXTENSION
CONTRACTOR "A" · BID SHEET –					
BASE COSTS - See Sheet 11			6000 –		3800 –
INS. & JOB COSTS (20% of LABOR)			760 – *		
OVERHEAD (5% of Mat. Cost)		5%	300 –	25%	950 –
			7060 –		4750 –
RETURN		5%	353 –	10%	475 –
SELL PRICE - MATERIAL			7413 –		5225 –
SELL PRICE - LABOR			5225 –		
			12638 –		
BID PRICE $12,600					

NOTE:
* INSURANCE AND JOB COSTS ARE LABOR BURDEN ITEMS BUT FOR MARKUP PURPOSES THEY ARE TREATED AS MERCHANDISING ITEMS

FIG. 22-1. Contractor *A* has a bid below normal because, being low on work, he has his better mechanics available and he feels justified in using low markups.

PRICING SHEET

DATE Nov. 1

JOB OR BLDG. DILLARD PRODUCTS **LOCATION** 1520 HIGH ST.
BID TO SAME

MATERIAL	QUANTITY	MATERIAL UNIT	EXTENSION	LABOR UNIT	EXTENSION
CONTRACTOR "B" - BID SHEET					
MATERIAL AND LABOR COSTS			6300 —		4200 —
ADD - MISC. LABOR					200 —
			6300 —		4400 —
JOB COSTS AND INS. (20% of labor)			880 — *		
OVERHEAD (10% of Mat. Cost)		10%	630 —	35%	1540 —
			7810 —		5940 —
RETURN		10%	781 —	10%	594 —
MAT. - SELL PRICE			8591 —		6534 —
LABOR - SELL PRICE			6534 —		
TOTAL			15125 —		
BID PRICE $15,125.					

NOTE:
* JOB COSTS AND INSURANCE ARE ITEMS OF LABOR BURDEN, BUT FOR MARK-UP PURPOSES ARE TREATED AS MATERIAL ITEMS.

FIG. 22-2. Contractor *B* has a bid above normal because his better mechanics are tied up and his office force is overloaded. Hence he must figure labor high and use high markups.

PRICING SHEET

DATE NOV. 1 19—

JOB OR BLDG. DILLARD PRODUCTS **LOCATION**
BID TO SAME **ADDRESS**
ARCHT. OR ENG'R. **ADDRESS**
PLANS MARKED **SCALE** **SPEC. No.** **EST. No.**
EST. BY **PRICED BY** **EXTENSIONS BY** **CHECKED BY** **S.O.No.** **SHEET No.**

MATERIAL	QUANTITY	MATERIAL UNIT	EXTENSION	LABOR UNIT	EXTENSION
∼NORMAL COSTS & MARKUPS∼					
BASE COSTS			6000 —		4000 —
INSURANCE & JOB COSTS (20% OF LABOR)			800 — ⌀		
OVERHEAD		10%	600 —	35%	1400 —
			7400 —		5400 —
RETURN		5%	370 —	10%	540 —
SELL PRICE - MATERIAL			7770 —		5940 —
SELL PRICE - LABOR			5940 —		
TOTAL SELL PRICE			13710 —		

EST. SELL PRICE $13,710.

NOTE:
⌀ — INSURANCE AND JOB COSTS ARE ITEMS OF LABOR BURDEN, BUT FOR MARKUPS, THEY ARE TREATED ITEMS OF MATERIAL.

FIG. 22-3. Normal bid for work figured by *A* and *B*. Note the methods of markup

The individual contractor must decide just how and when to use his reserve. Business may be good for the majority of contractors, but very poor for a limited few. Therefore, the contractor must regulate his markups according to his own business and not according to the general trend of business.

FAULTY MARKUPS

Readers of *Electrical Estimating* (McGraw-Hill Book Company, Inc.) are familiar with the effects of faulty and unbalanced markups. For those not familiar with the necessity of separate markups for material and labor, a brief explanation will be given.

On a dollar basis, it costs much more to supply labor services than it does to supply material. A study of the markups used in Fig. 22-3 will indicate the relative proportions.

On a job with a 60/40 M/L ratio, the final figure (neglecting insurances and job expense) will be the same with divided markups of 10 and 10% on material and 35 and 10% on labor, as it will with a common markup of 20 and 10% on both material and labor. On a job with labor in excess of 40%, a common markup produces a figure which is too low. With labor less than 40%, the final figure would be too high.

Contractors have a tendency to cling to the common markup method of pricing estimates. As long as they are figuring complete installation jobs with M/L ratios close to 60/40, the results are accurate. When they are figuring jobs with other ratios, their estimates are likely to be out of line.

A few years back, figures were taken on a two-story bus terminal and store building to be erected in Chicago's Loop district. There was much talk about the dangerously low price for which the electrical contractor took the job. I happened to be very familiar with the electrical requirements of this project. The purchase price of equipment and materials to be supplied greatly exceeded the estimated labor, and the price received for the installation was remunerative. The contractors contacted had used a common markup. No doubt, the successful contractor had figured his material and labor separately.

The recapitulations in Fig. 22-4 indicate the difference which common and separate markups can make on a final estimate. In this case, the project had an 80/20 M/L ratio (80% material and 20% labor). Use of a common markup produces a price which is $10,100 too high. Had the M/L ratio been 20/80, the price would be $9,100 too low.

				MATERIAL		LABOR	
	MATERIAL	QUANTITY	UNIT	EXTENSION	UNIT	EXTENSION
I	**ESTIMATE WITH SEPARATE MARKUPS**						
	CORRECT METHOD						
	BASE COSTS			80000 —		20000 —
	INSURANCE & JOB EXP. (20% OF LAB.)			4000 —		
	OVERHEAD — MAT. & LAB.		10%	8000 —	35%	7000 —
	ESTIMATED COSTS			92000 —		27000 —
	RETURN		5%	4600 —	10%	2700 —
	EST. SELL PRICE — MAT.			96600 —		29700 —
	EST. SELL PRICE — LAB.			29700 —		
	EST. SELL PRICE — THE JOB			126300 —		
II	**ESTIMATE WITH COMMON MARKUP**					
	WRONG METHOD					
	BASE COST — MATERIAL			80000 —		
	BASE COST — LABOR			20000 —		
	TOTAL BASE JOB COST			100000 —		
	INSURANCE & JOB EXP. (20% OF LAB.)			4000 —		
	OVERHEAD — 20% OF BASE COST			20000 —		
	TOTAL COST			124000 —		
	RETURN 10%			12400 —		
	EST. SELL PRICE — THE JOB			136400 —		
	METHOD II (WRONG) $136,400.					
	METHOD I (CORRECT) 126,300.					
	DIFFERENCE 10,100.					

PRICING SHEET — JOB OR BLDG. TERMINAL BLDG. — TYPES OF MARKUPS

FIG. 22-4. Both estimates have the same base costs ($80,000 material and $20,000 labor), but methods in markup make a $10,100 difference in the final bid price.

CONTRACTORS' RESOURCEFULNESS

Regularly we find the so-called wild bidder proving that he is just a little smarter than the other contractors on the bidding list. He can use tools and scaffolding to a better advantage, schedules his work better, selects his crews more carefully, makes arrangement with builders for clear floors, devises improved methods, and does any number of other things that keep costs down.

One weakness among many contractors is a desire to keep the door open for shopping around if they get the job. They do not want to be bound by commitments to any jobbers or equipment men. By doing this, they are closing the door against getting the job, because they do not get the better prices.

Contractors complain about vendors and vendors complain about contractors. We know that in many cases the complaints are justified. However, when they do get together, they both have a better chance of making a sale.

RUSHED ESTIMATES

Occasionally, the low bid is a faulty one resulting from rushed estimating. However, among the better estimators, bids rushed to completion are likely to be high. To play safe the estimator is generous with his allowances for quantities and contingencies and figures his labor somewhat above normal. The result is a final price which is appreciably above one that close estimating would produce.

Regardless of the causes, several high bids make the normal and correct bid look low.

STUDY LOW BIDS

Not all contractors discount their competitors' low bids. Instead many of them try to find some reason for their being low. They review their own estimate, and if satisfied it is right, they take other steps. If relations with the low bidder warrant, they contact him and compare estimates and notes.

Out of numerous jobs reviewed and studied, there may be only a small percentage that yield any worthwhile information. However, the material gleaned from the few may be extremely valuable. Besides, every contractor should take inventory of his methods and practices from time to time, and a wide spread in bids serves as a good tool to spur him into getting started.

CHAPTER 23

Safe Billing

DO NOT BILL YOURSELF OUT OF CUSTOMERS

One of the best assets a contractor has is a regular customer. It behooves him to slacken his pace occasionally in the race for new business and contemplate whether or not he is protecting and making the best of connections he already has.

The loss of a regular customer is of vital consequence. Could the loss have been avoided? The answer is often, yes. An outstanding cause for the loss of many customers is faulty billing. The statement was not comprehensible or was too high for the type of work billed.

Comprehensive Billing

In this chapter, we shall principally concern ourselves with billing according to the type and size of project. However, in passing, it is well to point out some of the features of comprehensive billing which have often been discussed by the author.

The following are some of the features of a comprehensible bill:

1. The direct job expenses are listed as such and not included as overhead.
2. Material markups are consistent with the cost of supplying same and are not designed to carry a part of the labor burden.
3. Labor markups are commensurate with the cost of supplying labor.
4. The markup for return (profit) is not excessive. It is not designed to cover shortage in overhead markup.

5. The breakdown reveals that the total is the aggregate of small charges.
6. The author of the statement makes sure that the buyer understands all the charges.

CORRECT BILLING

The contractor has two types of customers: (1) the regular buyer and (2) the occasional buyer. The regular buyer is the industrial or commercial institution that regularly supplies contracts without competition. The occasional buyer is the architect, engineer, or other type who keeps the contractor on his bidding list and awards a contract if and when the bid is inviting.

The Regular Buyer

It is the billing for the regular buyer that must be studied more carefully. There are no fixed rules for such billing, and the contractor must study each account so that he can create markups suitable to the contracts received.

Figure 23-1 shows representative overhead costs for contracts ranging from $100 to $9,000 and which comply with certain stipulated conditions. The values are for isolated jobs with an M/L ratio of 60/40 (60% material and 40% labor) and are obtained in competitive bidding. The regular buyer belongs in a different class and often rates better markups.

In the first place, work from the regular buyer does not entail the same expense as isolated projects obtained in the competitive field. The estimating cost is limited, and there is little or no sales expense. In the second place, orders are so frequent that often one job works right into the next, and at times, several small orders are, in the aggregate, equivalent to a large contract.

For study purposes, let us take a hypothetical case. We shall assume that a contractor has a customer who gives him 20 orders a year amounting to a total of $18,000. With 20 contracts amounting to $18,000, the average contract would be $900.

Figure 23-1 shows the normal job overhead for $900 to be approximately 24% (12% for material and 41% for labor). However, studies indicate that for the type and volume of business we are studying, an overhead for a job of two or three times the size of the average could be used. We shall use twice the average, which amounts to $1,800. Figure 23-1 shows the job overhead for $1,800 to be 21.5%

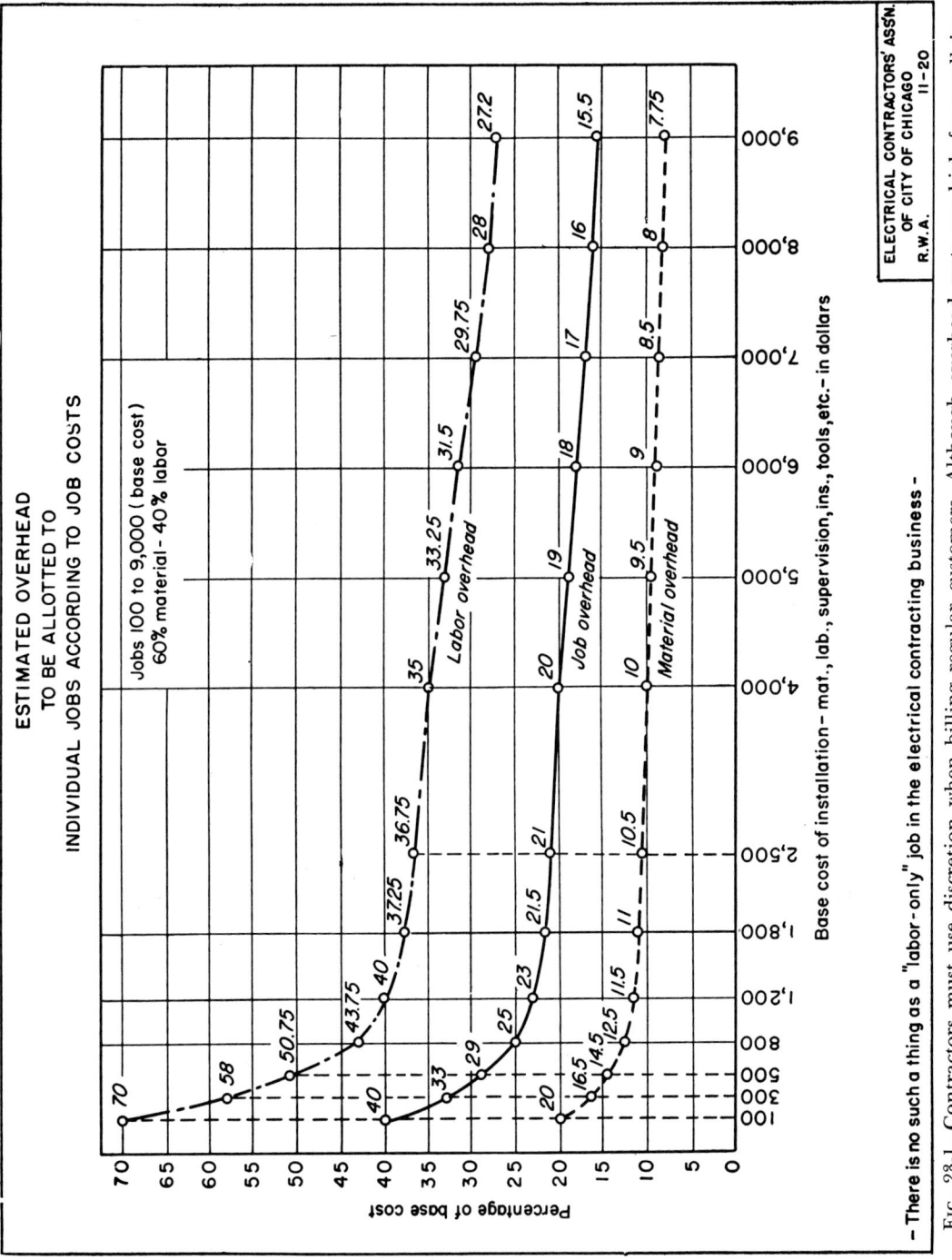

Fig. 23-1. Contractors must use discretion when billing regular customers. Although overhead costs are high for small isolated jobs, they may be much less for work from buyers who regularly supply work without competition.

(11% on material and 37.25% on labor). If twice the average contract is used, the job overhead is reduced from 24% to 21.5%.

Contractors like to stay away from odd percentages. In the case we have been studying, chances are that under favorable conditions, a con-

A STUDY OF OVERHEAD COSTS

For Billing Regular Customers

Order numbers	Approx. billing (1)	Normal % (2)	Normal Dollars (3)	Reduced * % (4)	Reduced * Dollars (5)
754	$ 800	25	200	22	176
755	400	30	120	25	100
756	2,800	20.5	754	18.5	518
757	450	30	135	24	108
758	700	26	182	22	154
759	650	27	176	22.5	146
760	600	28	168	23	138
761	4,600	19.5	897	15.4	708
762	600	28	168	23	138
763	200	35	70	30	60
764	400	30	120	25	100
765	300	33	99	28	84
766	450	30	135	24	108
767	300	33	99	28	84
768	400	30	120	25	100
769	450	30	135	24	108
770	500	29	145	23.5	118
771	400	30	120	25	100
772	800	25	200	22	176
773	2,200	21	462	19.8	436
Totals	$18,000		4,505		3,660

* Overhead used was for job twice as large as shown in Col. 1.

FIG. 23-2. Reduced overhead markups can be used for regular customers supplying work without competition. In the example shown, the reduced overhead saves the buyer almost 5% of the job cost.

tractor would use 10% for material and 35% for labor. With a 60/40 M/L ratio, the resultant job overhead markup would be 20% instead of 21.5% as shown on the curve.

Figure 23-2 shows the effect of applying the reduced overhead to the

Safe Billing

20 contracts amounting to $18,000. Column 2 lists the normal overhead (for isolated jobs) and Col. 4 lists the overhead for jobs twice the actual size. For example, the first contract is for $800 and the 25% in Col. 2 is the normal overhead for that amount. The 22% listed in Col. 4, on the same line, is the normal overhead for twice that amount, or $1,600.

The totals in Fig. 23-2 are $4,505 for the normal and $3,660 for the reduced overheads. The former represents approximately 25% and the latter slightly over 20% of $18,000.

In practice, the contractor would more than likely want to use the same markup for all billing. If the reduced overhead for the average job ($1,800 overhead for $900) were used, the amount would be 21.5%. Under very favorable conditions 20% (10% on material and 35% on labor) would be used.

From the foregoing we get the following figures:

1. Normal overhead applied to each job
   (Col. 2, Fig. 23-2) $4,505
2. Reduced overhead applied to each job
   (Col. 4, Fig. 23-2) $3,660
3. Special overhead of 20%
   (10% on material and 35% on labor) $3,600

In a story the figures are not startling enough to impress one. However, in practice they are more effective.

To augment the foregoing, two types of billing can be prepared for the average $900 (base-cost) job. The first will be with a common markup based on normal experienced overhead, and the second will be with adjusted individual markups. The normal overhead (job) for $900 was listed as 24%. The contractor would use 25%. Our final figure for the special overhead was 20% (10% on material and 35% on labor).

The first example gives the impression that material is carrying a burden of 25 and 10%. It does not indicate that the 25% is not sufficient to cover labor costs.

The second example shows true costs. There are two reasons why this should be done:

1. The true charges can be substantiated
2. Experience has taught us that true markups (low on material and high on labor) get a better acceptance than a common markup designed to cover the total job cost.

It is to be remembered that each contractor must establish his own experienced overhead costs. Studies such as are shown in Figs. 23-1 and 23-2 serve as guides only. A contractor's studies must be complete. A little knowledge of operating costs has often proved dangerous. Let us review an actual case.

Example 1. Normal Overhead and Common Markup

Material (60%)............	$ 540.00
Labor (40%).............	360.00
	$900.00
Overhead (25%)..........	225.00
	$1,125.00
Profit (10%).............	112.50
Total...................	$1,237.50

Example 2. Adjusted Overhead and Separate Markups

Material (60%)............	$540.00	
Overhead (material) (10%)..	54.00	
	$594.00	
Return (5%)..............	29.70	
Total material...........		$623.70
Labor (40%).............	$360.00	
Overhead (labor) (35%)....	126.00	
	$486.00	
Return (10%).............	48.60	
Total labor..............		534.60
Total the job...........		$1,158.30

FIG. 23-3.

A SPECIFIC CASE

As stated before, studies do not impress us. However, some actual experiences have been startling. An outstanding case is that of a Wisconsin contractor who lost some of his better industrial accounts after suddenly becoming overhead-conscious.

For a long time the contractor had used established markups for the installation portion of his business. The exact figures are not known, but they were approximately 15 and 10%. The 15% was supposed to be overhead, and the 10% was for profit.

One year, the contractor decided to make what he thought was a careful study of his overhead costs. The conclusion was that the experi-

Safe Billing

enced overhead was 40%. Immediately he began adding 40 and 10% to all bills regardless of the type or size of job being billed. He insisted that it cost him that much to operate and he must add that much to the base cost of the job. Soon, customers began dropping off.

There were several things wrong with the contractor's figuring:

1. The 40% was not all overhead. It included many direct job expenses, some of which had already been charged to the work.
2. The 40% included the upkeep of a small store and repair shop and the overhead for many small maintenance and repair jobs.
3. The markup of 15 and 10% which he had been using was misleading. The true figures would have been more like 21% for overhead and 5% for profit.
4. No consideration was given to the reduced cost of serving regular customers who supplied large volumes of noncompetitive work.

Other cases have come to our attention where contractors had billed themselves out of customers. The billing may not have been excessive, but the method was wrong, and the buyer did not understand that it was just.

THE DEMORALIZING EFFECT

The morale blow that one receives from the loss of a customer is often more serious than that received from the financial loss. One hates to see a competitor take away his regular accounts, and if it does happen, every time he thinks of the competitor or the customer, he is reminded of the loss. Financial losses on jobs and the loss of contracts in competition may soon be forgotten, but the memory of a good account lost lingers for a long time.

We not only regard customers as a source of business but often consider them good friends. When we lose a man as a customer, we feel that we have lost a friend as well.

THE OCCASIONAL BUYER

We have classed architects and engineers as occasional buyers. Orders received from them are usually fixed-price contracts for the main project, but extras follow. If the extra orders are given on a cost-plus basis, the billing requires the same care as that for regular customers. There is much more to be said about the architects and engineers, but dealing with them is a subject to be treated separately.

IN GENERAL

There are salesmen, manufacturers, mechanics, and others who feel that their personal interests can be best served if they can persuade owners to buy materials directly and hire mechanics to install them. Where suitable electrical contractors are available, there are few if any cases where the owners can benefit by trying to do their own work. Nevertheless, the hazard is there and the contractor must do all he can to combat it. Careful billing is one of his most effective weapons.

It would be gratifying if a contractor's honest efforts were always fully rewarded, but we know that such is not the case. In spite of all that one can do, changing times and changing conditions will bring about losses. However, we know that the greater majority of better contractors have built up their business by careful handling of customers. The best one can do is to plan his operations carefully and be on his guard. If losses do occur, he will not feel at fault.

CHAPTER 24

Architects and Engineers

In his daily negotiations with architects and engineers, the electrical contractor takes it for granted that his methods cannot be improved. A review of many experiences gives us reasons to think otherwise and conclude that the following justify study:

1. Methods of presenting propositions
2. Figuring and billing extra work
3. Recommending changes in plans and specifications
4. Protecting the architect
5. Protecting the owner

PROPOSITIONS

In Chap. 13, propositions for electrical work were discussed at length. We learned then that in presenting the main bid to architects or engineers, little was required other than the statement that work was to be done in accordance with the plans and specifications. Later in this chapter we shall learn that bidding for extra work can become more involved. Under the heading "Recommended Changes," we shall also learn of things to be avoided when writing propositions.

ESTIMATING AND BILLING

There are three principal methods of estimating and billing for additional work, namely:

1. In accordance with unit prices stipulated in the main contract
2. At cost plus percentages of markup as stipulated in the main contract
3. In accordance with the contractor's regular practice (no provision in the contract for method of billing)

UNIT PRICES

In their specifications, architects and engineers frequently require a list of unit prices to be used in figuring the cost of any additional work that may be needed. These prices are to be included as a part of the main bid. The units are so much per foot of conduit and wire and so much each for outlet boxes, switches, etc.

The architects justify their request for unit costs on the grounds that they facilitate checking and provide protection for the owner. Experience leads us to question both of these reasons as being generally true. Later we shall study another method which is just as fair to the owner, as simple for checking, and relieves the contractor of the burden imposed by a unit price for each and every item of material that may be required for additional work.

The Contractor's Expense

It is costly enough to figure work against competition without having to shoulder the extra burden of figuring and quoting on unit prices. A contractor may have prices at hand, but conditions vary so much that no set of units are generally applicable. Types and sizes of jobs are not uniform, and prices change frequently. By any method used, the preparation of unit prices is costly.

There are two general methods of preparing unit costs: (1) figuring the individual unit, (2) establishing a multiplier to be applied to all materials belonging in a similar group.

In Example 1, there are 16 (not including figuring discounts on material and extending labor) operations for a single item. The method illustrated by Example 1 is generally used for motors, starters, panels, and other sizable pieces of equipment. It is common practice to arrange the figures in a horizontal line with a column for each operation. The arrangement for the material section of the sheet is approximately as follows:

Col. 1. Item to be priced
Col. 2. List price of item
Col. 3. Discount to be applied
Col. 4. Purchase price

Architects and Engineers

Example 1. The Individual Unit Method for 1,000 Ft of No. 000 RHRW Wire

	Material		Labor
Base costs	$550.00	(45 hr @ $3.50)	$157.50
Shrinkage (material) (5%)	27.50		
Insurances (labor)	.....	(12%)	18.90
Miscellaneous and direct job expense (2%)	11.00	(6%)	9.45
	$588.50		$185.85
Overhead (10%)	58.85	(35%) Base lab.*	55.12
	$647.35		$240.97
Return (5%)	32.37	(10%)	24.10
Sell material	$679.72		$265.07
Sell labor	265.07		
Total sell price per 1,000 ft	$944.79		
Sell unit per foot	$0.94479	(use $0.95)	

* The 35% is not added to insurances and the 6% job expenses.

Col. 5. Miscellaneous and direct job expense (this may require separate columns for cartage, hoisting, tools, etc.)
Col. 6. Total job cost for material (Col. 4 plus Col. 5)
Col. 7. Overhead markup *
Col. 8. Total (Col. 6 plus Col. 7)
Col. 9. Markup for return *
Col. 10. Sell price (Col. 8 plus Col. 9)

Ten columns have been used for material. There will be as many or more for labor. At times, as many as 30 columns are used.

Example 2. The Common Multiplier

	Material, %		Labor, %
Base cost	100		100
Allow for shrinkage (material)	5		
Insurances (labor)	....		12
Miscellaneous and direct job expenses	2		6
	107		118
Overhead markup (10%)	10.7	(35%) Base labor *	35
	117.7		153
Markup for return (5%)	5.89	(10%)	15.3
Totals	123.59		168.3
Multiplier for material	1.2359	(use 1.24)	
Multiplier for labor	1.683	(use 1.70)	

* The 35% markup is not applied to insurances and job expenses.

* Overhead and return columns usually have two divisions: one for percentage and one for dollars.

Applying the results of Example 2 to the base costs ($550 for material and $157 for labor) of No. 000 RHRW wire, we have:

$$\text{Material cost per 1,000 ft ($550} \times 1.24) \ldots \ldots \$682$$
$$\text{Labor cost per 1,000 ft ($157} \times 1.70) \ldots \ldots \ \underline{268}$$
$$\text{Total unit price (sell) per 1,000 ft.} \ldots \$950$$
$$\text{Unit price per foot} \quad \$0.95$$

The multipliers are used for conduit, wire, outlet boxes, and other materials that are not likely to involve special engineering or handling expenses.

The simplified method of using multipliers still involves several calculations. Prices must be checked, discounts applied, and multipliers used. There are the separate calculations for the material and labor. The two results must be added, and the final check made. After all the calculations are completed, the prices must be written into the proposal.

SPECIFIED MARKUPS

A contract with stipulated markups and clear-cut statements of what is to be included as direct job expenses simplifies estimating. Although the contract simplifies estimating and provides for easy checking, it has one serious fault. It imposes the same markups on sizable contracts as it provides for the small nuisance jobs.

VARIED MARKUPS

An outstanding cause for the unit price system has been the belief that contractors overcharged for additions and change orders. There may be some justification for such beliefs, but we do know that in many cases, the extra or change order is little more than a nuisance as far as the contractor is concerned. Many of us have long been of the opinion that projects completed with few or no additional orders are far more desirable than those involving frequent changes and additions.

A common reason for buyers being skeptical about estimates for extra work is the mistaken idea of many contractors that buyers have no right to know how the work is figured. The architect or engineer is given a price and expected to accept it without knowing the detailed costs.

The better contractors have long since learned that they have little or nothing to lose and much to gain by taking their estimates to the buyer and acquainting him with the detailed costs.

Barriers against acquainting buyers with estimating details are the high costs (labor units and markups) of small orders. Varied markups and educational work can be used to lessen the seriousness of these obstacles.

Submitting Varied Markups

Slaves of habit that we are, there is a tendency to go on as we have in the past. We do nothing about change orders until the work is at hand. Then we submit a price and hope it will be approved. Why not go to the buyer as soon as changes are contemplated and explain that costs vary greatly depending on the amount of work involved. A list of markups for material and labor, similar to the following, could be submitted:

Dollar Volume	Markups, %	
Base cost of work	Material	Labor
100– 300	20 and 10	60 and 10
301– 800	15 and 10	50 and 10
801–1,200	12 and 8	42 and 10
1,201–2,000	11 and 6	36 and 10
2,001–3,000	10 and 5	35 and 10

To augment the list of markups, base costs such as $18 per 100 ft of ½-in. conduit and $20 per 1,000 ft of No. 14 wire could be supplied. These figures would serve as a guide for the costs of all sizes of wire and conduit. Similar base costs could be supplied for other wiring materials.

In a previous chapter, under "Safe Billing," the subject of selecting markups to suit certain conditions when dealing with regular customers was treated at length. At times, similar precautions may be in order while dealing with architects and engineers. Again, it may be a case of several small orders amounting to the equivalent of one sizable contract.

The architect or engineer may prefer a uniform markup for all work. Nevertheless, submitting the suggested list of sliding markups would pave the way for an amicable settlement. The buyer would realize that the cost of small orders was high.

There is a great tendency to vary markups in units of 5%. Markups such as 15, 20, 25%, etc., for overhead and 5, 10, and 15% for profit are used. How absurd it is, after striving so hard to keep the base costs accurate, to jump around with 5% at a time for markups.

It will be noted that such values as 6, 11, and 42% are used in the foregoing example of sliding markups. They are used because they are right and will help sell the idea.

If a contractor can avoid unit prices by getting a list of markups approved, it is something accomplished. It will lessen his work, cause no hardship to the architect, and protect the owner.

EDUCATIONAL WORK

As the author has often stated, opportunities for educational work are limited and the contractor must bide his time. When negotiating contracts and orders for additional work, one can easily get an audience. At such a time the architects and engineers are interested in a contractor's costs and there is a specific project to study.

Without going into details, let us list some of the items responsible for the high cost of small orders which should be brought to the attention of customers.

1. Time required to get information regarding the work
2. Estimating time out of proportion to dollar cost
3. Cost of getting extra or special material to the job
4. Cost of supervision
5. Interruption of other work for special jobs
6. Regular work delayed while deciding on changes
7. Cost of bookkeeping and billing
8. Cost of supplying electrical installations much higher than that for other trades

Statements regarding such costs are much more effective if substantiated by published data.

RECOMMENDING CHANGES

Architects and engineers prepare plans and specifications which represent their idea of what is best for the owner. A contractor wishing to recommend changes must use discretion.

In writing propositions one must avoid suggested changes in the plans without the approval of their author. And if changes are suggested in a proposition, it should be stated that they are made with the approval of the author.

One should keep in mind that an owner usually sees the contractor's propositions. If the owner sees suggested changes, he may get the idea that his architect has not specified the most suitable installation.

As a rule it is well to approach the author of plans and specifications with questions such as the following: Have you ever considered this . . . ? Do you think that there is any chance that the owner will want this or that? Will you give me a check on my figures?

Just how far one goes with recommending changes depends on his relations with the architect. Architects have a high regard for some contractors and welcome their check on plans and specifications. When estimating a job, a contractor can often detect omissions or shortages that are easily overlooked at the time the plans are given a final check in the architect's office.

The Owner

After a contract is under way or when figuring alteration work, a contractor cannot avoid contact with the owner. At such times he must be careful about recommendations. It is so easy to answer an owner's questions and then realize afterward that the architect should have been consulted first.

When the contractor is contacted by the owner, diplomacy is again required. A good stock answer to an owner's questions is, "You had better talk to the architect about that. I do not know just what he had in mind."

PROTECTING THE ARCHITECT

One does not like to suggest changes, but at times it is imperative. Items specified or shown may not comply with the code, feeders or panels may not be up to capacity, or items may have been entirely overlooked. If the architect's attention is not called to such things when the work is being figured, the mistakes will prove costly later on. Besides, the architect will find it embarrassing when the owner learns of the shortcomings of the plans and specifications.

As stated before, when a job is being figured, it is easy to detect shortages. For this reason, many architects look to the reliable contractors for a final check on the electrical work. Close relations between architects and contractors benefit both and safeguard the owner.

PROTECTING THE OWNER

Much in this chapter has noted the architect's efforts to protect the owner. We have also seen how the contractor can cooperate with the architect for the same purpose. Too much emphasis cannot be placed on the necessity of all trades and the architect working together so that the original contracts will provide a complete installation and extra orders can be avoided.

Many owners have had the exasperating experience of completing their projects at costs greatly in excess of the amounts of the original contracts. In most cases, the fault was their own for not having supplied the architect with the proper information or because they had changed many things after the work was under way. Nevertheless, they felt that the contractors had charged them excessive prices and that the architect was partly at fault.

Some professional builders hire their own architects and engineers and provide a complete building service. They get much business because they stress the fact that they will give a guaranteed price for the completed project. Their guarantee is no different from that which has long been supplied by the architects. They guarantee a price for a building including all work covered by the plans and specifications. There is no guaranteed price covering any additions or changes that the owner might wish to make after the original plans are approved.

In spite of all that can be done, owners will ask for changes and additions after the main project is under construction. Often they will be displeased with the costs of these. The following are "must" items for keeping building costs in line and maintaining the confidence of the buyer:

1. Architects must make a special effort to get complete information regarding the owner's requirements.
2. Architects must impress the owner with the danger of excess costs if pertinent information is not forthcoming before contracts are let.
3. When estimating, each trade must make every effort to note any shortcomings of plans or the possibility of future needs.
4. Each trade must guarantee a complete job for the contract price.
5. Each trade must be prepared to absorb the costs of some minor changes and additions.

With a good architect, capable contractors, and the above practices, the owner will be well protected.

MAINTAINING GOOD WILL

In this chapter, some practices in selling, such as making contacts, cultivating friendships, the country club, and the golf course, have not been discussed. Here we have been interested in maintaining the good will and caring for the needs of existing clients.

There was a time when the electrical contractor might have been excused for leaving the architect and engineer in doubt as to the fairness of his charges, but in these times, with the advanced knowledge of costs and the great amount of research data available, there is no reason why there should be any quarrel about prices.

The contractor's chief business with architects and engineers is selling electrical installations. If he is to maintain their good will, he must convince them that his prices are right. He has the source of knowledge and the data available to do this.

CHAPTER 25

The Estimator

As business grows, in order to avoid long hours of overtime, additional help must be employed. In studying Fig. 2-1, Division of proprietor's time, we readily see that for a one-man business, the first logical addition of help would be an estimating engineer. The table shows 25% of the proprietor's time allotted to estimating and 10% to engineering. A capable estimator could take over these duties and help with others such as making contacts, purchasing, and billing.

Before we enter into the discussion, we must establish limitations for the meaning of the word *estimator*. In the trade, it refers to men engaged in a wide range of work. It may be applied to the man in the back room grinding out quantity take-offs, or it may refer to a man in the front office who not only prepares complete estimates but is active in all phases of the business.

For our purpose, the minimum requirement for an estimator will be one who can prepare a complete estimate from plan study to final bid price and can supply such engineering as is required of the firm for which he works.

Electrical Estimating (McGraw-Hill Book Company, Inc.) provides details regarding the selection and training of electrical estimators. In this connection the discussion will be limited to questions to be decided by management, such as the following:

Freedom of activity
Responsibility

Overtime and overtime pay
Getting bids out on time
Contentment

Not all readers are aware of the conditions under which some estimators must work. Hence, certain ones will be surprised that the author finds some of the following comments necessary.

We assume that our estimators are high-grade men with good judgment and honest. It goes without saying that such men will always try to act in accordance with the wishes of management.

FREEDOM OF ACTIVITY

To do his best work, the estimator must feel free to contact outside sources for information and prices. He must also feel free to contact the men in the field if in the best interests of the work. Where there is more than one estimator, jealousy on the part of the senior estimator often imposes restrictions on the newer men.

A study of costs is an essential part of an estimator's work. To engage in this, he must be permitted to follow the costs of his jobs while in progress. He must also have time to study new construction methods and keep his labor-cost units up to date.

The contractor who tries to confine his estimator's work to narrow limits usually gets narrow results and often ends up by losing good men.

The subject of labor-cost studies is treated at length in previous chapters.

RESPONSIBILITY

A good estimator understands that he is strictly responsible for getting out complete, accurate estimates in a reasonable length of time. Five per cent is the normal tolerance for estimated material and labor costs. Any marked deviation, up or down, in the final costs must be explained. It is an exacting condition, but a part of the trade.

Conditions and contractors vary the extent to which the estimator enters into the actual construction of the job. Regardless of the limits of his activities, he has a responsibility from the time the estimating plans are unrolled until the last piece of material is installed.

It is the constant prod of responsibility that spurs men on to do their best work, and it is the credit for work well done that lessens the strain of their toil.

OVERTIME AND OVERTIME PAY

Every estimator expects to work extra hours at times. Electrical contracting demands it. However, overtime must be avoided as much as possible for three principal reasons:

1. Eight hours is close to the limit of time for a good day's work, week in and week out.
2. Overworked men cannot supply the push and speed required to get rush jobs out on time.
3. Tired estimators slow down and are liable to make serious mistakes.

Management must see that the flow of work to be estimated is as uniform as possible and that there are enough men to avoid excess overtime. There must be times when there is a shortage of work to figure, so that men will have time to take care of job studies and other essential incidental work.

OVERTIME PAY

There is no uniform practice among electrical contractors regarding extra pay for overtime. Some of the better contractors have never given the subject a thought. They pay reasonable salaries, and their estimators expect to work some extra hours in case of emergency.

There are arguments for and against the overtime-pay practice. However, arguments that apply to one contractor-estimator combination will not apply to another. Pages could be written on the subject because to cover it, one would have to deal with no end of personalities and conditions. We must confine our discussion to a limited number of observations.

Once the practice of paying overtime is established, it is hard to change. Any contractor contemplating it must consider the following questions as they apply to his individual contractor-estimator combination:

1. Will it tend to create a barrier between the estimator and management?
2. Will there be a tendency to overwork estimators because management feels that it is not asking any favors when requiring extra hours?
3. Will any estimators take it to be an indication that they are not expected to get their work out in an 8-hr day?

4. Is the estimator receiving overtime pay as contented as the straight salaried man?

To this list more questions could be added, but this is enough to show that the question requires study.

Some readers will ask, "Why wouldn't a man be happier with pay for overtime?"

We assume that in the end, both men receive much the same pay. The straight salaried man knows that he owes some overtime to the business and gives it in the spirit of cooperation. His salary being fixed, he and his family can plan their budget accordingly.

The man who receives overtime pay one week wants more the next. If overtime drops, he feels underpaid. The foregoing questions imply that overtime pay can place the estimator in an undesirable position.

It is the author's opinion that with the normal contractor-estimator combination, both management and estimator are best served with the straight salary arrangement.

GETTING BIDS OUT ON TIME

When being retained to study over-all business operations, the author has had contractors complain that at times their estimating departments did not have estimates ready on schedule. It was usually no fault of the estimators.

In the best managed offices, there are times when bids cannot be ready according to schedule. There are two principal reasons for this:
   1. Time allowed for getting in bids is unreasonable.
   2. Contractors cannot afford to keep enough estimators to meet all peak demands.

Contractors must be on the lookout for unreasonable requests. As soon as they learn that not enough time has been allowed for preparing the estimate, the owner must be notified. There may be no immediate extension of time, but as the bidding date approaches, the owner will learn that contractors must have a reasonable amount of time for bidding.

Contractors must learn that there will be times when owners will have to wait to get a price. However, the number of times can be held to a minimum by careful management. The office must avoid getting overloaded with work that has little to offer. Work should not be brought in just to keep estimators busy. They do not need the practice, and their time can be spent to better advantage doing other things. Besides, good work may come in unexpectedly.

Important projects should be started promptly. An estimator may think there is lots of time, but he should be constantly reminded of that universal fault of underestimating the time it takes to do things. Two or three days may slip by and not seem like much before a big job is started, but a single day is a lot of time just before the bid is due.

CONTENTMENT

We know that a man's state of contentment has a great deal to do with the amount and quality of work he turns out. Contentment is more vital to estimating than many other lines of work. The estimator to be at his best must be contented with his salary, surroundings, work, associates, and all else. He must be happy not only at the office but in his home life as well.

Management cannot enter into the direction of one's home life, but at times it can help promote happiness and avoid anxiety. It can make an employee feel free to take time off in case of sickness at home or pressing personal matters. It can also see to it that he has his vacation to spend with his family and is not tied down to the office so much that it interferes with his home life.

A contractor owes a loyal estimator more than just a salary; he owes his personal welfare every normal consideration.

CHAPTER 26

Man-power Productivity

A study of man-power productivity (MPP) is everyday practice in electrical contracting. For every job that is figured, labor-cost units must be adjusted according to the anticipated productivity of the men to be employed. Normally this estimating of MPP is not too elaborate a process. The estimator knows his mechanics and has similar completed projects to use as a gauge.

However, there are times when estimating MPP is not simple. The project is unusual, many new men will be required, and the labor situation is bad. It is then that one is required to be familiar with all the factors that affect productivity, and knowing these factors, one must still know how they themselves vary with conditions.

We know certain facts regarding MPP, and around these we must fabricate figures for our estimating purposes. We know that the three outstanding factors to be considered are:

1. The labor market (mechanics available)
2. Management (support of the men in the field)
3. Working conditions

Each of these items has several subdivisions.

THE LABOR MARKET

When we speak of the labor market we are referring to the mechanics available. Therefore we must study the qualifications of the individual.

The Individual

The principal factors effecting the individual's ability to produce are:

Item	Remarks
Physical qualifications	Depends on type of work
Education and training	Including on-job experience
Natural propensity	Skill and initiative
Attitude	Desire to do well
Disposition	Ability to work with others
Home life	Including off-the-job activities

Of the foregoing list, there are only two items that are not readily understood by the reader—home life and attitude. Home life must be explained and attitude elaborated on.

Home Life. To produce his best work, a man must have a restful and happy home life. Men who are not happy are liable to slacken their pace, be irritable, and make mistakes.

Some mechanics try to build houses, work at service stations, or do any number of other things while working for contractors. As a result their regular work suffers.

Electrical construction does not prove inviting to bums. Hence, in normal times, contractors have little trouble because their men are up late at night gambling or hanging around taverns.

Attitude of Mechanics. When there is a shortage of man power, some of the less desirable mechanics must be tolerated. When we say tolerated, we are thinking of the men with the wrong attitude.

There are many characteristics that make one mechanic less desirable than another. Studies show that the wrong attitude is outstanding.

The effect of the attitude of a mechanic on productivity has never entirely escaped the attention of electrical contractors. Its full significance, however, is not realized until we begin to spell it out with figures.

The accompanying table, Fig. 26-1, provides a comparison which points out the ill effects of a poor attitude. Column A represents the record of an indifferent mechanic, and Col. B lists the author's proposed standards which are based on field experiences of well-managed projects. To understand the listings thoroughly, some discussion is necessary.

A STUDY OF COLUMN B, FIG. 26-1

General

Before studying the detailed figures in Col. *B* Fig. 26-1, let us look at the total time. The total is 80 min, or 1 hr and 20 min. Over the years, contractors have found that 1 hr is ample time for the items listed, not including the "coffee break." The coffee break accounts for the extra 20 min in the total for Col. *B*.

Changing Clothes

There are differences of opinions regarding time allowance for changing clothes in the morning. Without going into the pros and cons, let us again look at the total time of 80 min. That is all a job can be expected to stand for such phases of incidental labor.

At the present high rate of pay, there is a limit to how much the buyer should be expected to pay men for "not working." Besides, the contractors' customers expect their men to be dressed and at their machines when the starting whistle blows.

Selecting Tools and Getting on the Job

The tools referred to are ordinary hand tools. Special allowance must be made for large pieces of equipment such as pipe benders, cable pullers, power saws, etc.

Here we have 10 min to select tools and get to the location of the work. This is ample for an ordinary industrial project. The allowance must be greater for multistory buildings, power plants, and projects where the work is spread out.

Many buyers think that because they are paying for 8 hr a day, the men should be out on the job installing work for 8 hr. This contractors know is not practical.

Coffee Break

The formal coffee break is of recent origin, and many of us are of the opinion that the mechanics are entitled to it. Besides affording the men a brief and pleasant respite, it often pays dividends. It stimulates, relieves tension, and builds up morale.

On large projects, arrangements can be made to have fresh hot coffee delivered right to the men. If it is not convenient for management to supply coffee, men must bring their own.

Very often, on small projects, the coffee break is prolonged but the time is not wasted. While the men have time out, they talk over and plan their work.

Lunch Periods

Thirty minutes out for lunch is established practice, and the advisibility of a longer period will not be discussed. With this short period, additional time must be allowed for washing up and getting to and from the point of work. The allowance for this is not great, but along with other allowances, it adds up.

Other Listings

The reasons for time allotted to relief periods and pickup are obvious. Again we must remind ourselves that a special type of project is being studied. Under many conditions the allowance would have to be greater.

Under any conditions, toilet facilities must be adequate if the 80-min schedule is to be maintained.

A STUDY OF COLUMN A, FIG. 26-1

Having established the values in Col. *B* as reasonable, we know that those in Col. *A* are excessive. Looking at Col. *A*, we do not find any values that seem radically higher than those in Col. *B*, yet the little here and there, added to the "Lack of despatch" item, gives a sizable total.

Column *A* has a total 112 min greater than Col. *B*. That is almost 2 hr, or 25% of an 8-hr day. Twenty-five per cent is a large excess load to be put on any job.

We have been studying a special case to show what a terrible hazard the man with the wrong attitude is. Fortunately it is not representative of the average mechanic.

Groups of men working on small or medium-size jobs are not so large but what the contractor or his foreman can get a very good idea of the values of the individual workmen. If other workmen are available, the poor ones can be weeded out. When the labor market is bad and contractors cannot sift their gangs too carefully, the poorer mechanics may be assigned to groups where they will have to work to keep up. There are lots of good mechanics who will make going very rough for the fellow that does not want to keep up his end of the work.

There are mechanics of mediocre ability who, working alone or with someone they do not like, do not do well. These same men when assigned with someone who understands them can produce an average or better than average day's work.

ATTITUDE OF MECHANICS
Effect on Productivity

Operation	Time, Min (See Notes) A	B
Changing clothes in morning	10	
Selecting tools and getting on the job	15	10
Coffee break—9:30 A.M.	15	10
Preparation for lunch	15	10
Return to work after lunch	15	5
Coffee break—2:30 P.M.	15	10
Relief periods	30	20
Pickup and prepare to leave—afternoon	20	15
Sitting around waiting for quitting time	10	
Carelessness (false moves, spoiled work, etc.)	15	
Totals	160	80
Lack of despatch—10% of 320 (480 − 160)	32	
Totals	192	80

Excess time for A $(192 - 80) = 112$ min

Notes:
 1. Values are for an ordinary industrial job.
 2. Column A represents record of an indifferent mechanic.
 3. Column B lists proposed standards.

FIG. 26-1.

Hiring New Men

The layman may wonder why the electrical contractor does not go through the formality of having men fill out applications and furnish letters of recommendation. It just is not done with the building trades. In the first place, the mechanic thinks his union card is evidence enough of his ability and resents being asked for letters. In the second place, most contractors prefer to get their information firsthand.

Individual Productivity

There is no exact standard for measuring the productivity of the individual. A long story could be written on the subject, but to simplify our work, let us decide on the following:

During normal times about 90% of the available electricians are employed. We shall say that the average productivity of this group represents 100% for the individual. In the study of group or crew productivity, we shall contemplate the conditions permitting such results. A man not only must have the ability to produce 100% but must be favored with the support and working conditions that permit it.

CREW PRODUCTIVITY

In order to present all problems of MPP we must have some standards. Normal productivity and conditions affording it must be decided.

We shall say that normal productivity is 100% and that this value is realized when the following conditions exist:

1. Labor market—only 90% of the available mechanics employed
2. Type of work—ordinary industrial
3. Material deliveries—good
4. Support of men in fields—good
5. Working conditions—favorable
6. Hazards—allow 10% for variations from conditions stipulated

The first five items listed represent close to ideal conditions which rarely, if ever, exist. There are many hazards of material deliveries, field support, working conditions, etc., which can consume the 10% added.

Not all mechanics will have the same rating. Let us say that out of every hundred men, we have the following groups with ratings as listed:

No. of Men	Product Rating, %	Extension
30	110	3,300
40	100	4,000
30	90	2,700
Totals 100		10,000

Average rating 100%

Out of the 100 men there are 30 with ratings 10% above normal and 30 with ratings 10% below normal. In normal times and with standard conditions one may expect the MPP of crews to range from 90 to 110%.

Having studied the labor market and the ability of mechanics to produce, factors that influence the mechanics' ability must be studied.

MANAGEMENT

In previous chapters we learned the importance of having management support the men in the field by supplying needed information, working drawings, and adequate tools. It goes without saying that the proper selection and placement of men is also a function of management.

Contracts must be selected in accordance with the labor market and the limitations of the contractor's organization. When the MPP on a job is bad, the first question is usually, "What is wrong with labor?"

The first question must be, "Is it labor or management that is at fault?"

WORKING CONDITIONS

A list of working conditions which affect productivity will prove enlightening. The following are some of the principal factors which must be contemplated:

1. Condition of working spaces
2. Material deliveries
3. Cooperation of other trades
4. Progress of the job as a whole

The reader has already met with most of these factors in various ways. In Chaps. 7, 8, and 9, their effects were viewed from different angles. Here phases of progress beyond the control of the electrical contractor must be studied.

Progress of the Job

We hear much about the hazards of bidding on the electrical contract before the general contractor has been selected. The progress of the job as a whole depends on how well the general contractor manages his work. Irregular progress of the work and duration of contract in excess of normal impose heavy costs on the electrical contractor. He has the extra cost of men on and off the job or standby labor. The extended duration also puts a burden on the entire organization (see *Electrical Estimating*, McGraw-Hill Book Company, Inc.).

For some types of construction, weather may be a factor affecting job progress. On outdoor underground work, it is a hazard throughout the life of the job. On a multistory slab-constructed building, it is a hazard only until the building is under cover.

The work of architects who let the contracts for all trades directly and do their own superintending is much sought after. They select their contractors and manage their work well. As a result, progress is steady and completion dates timely.

ADJUSTING LABOR UNITS

The novice will be surprised to learn that the majority of estimators go along day after day adjusting labor-cost units without writing down any figures. In the first place, by studying such problems as we have treated here and frequently discussing them with others, they become familiar with the effect of varying conditions. In the second place, much of their work is more or less uniform and required adjustments not radical. Often, only labor costs for certain items need be adjusted.

There are times when it is well to set down some figures. For example, let us take a project requiring 40 men with conditions depicted by the following:

Productivity of mechanics available:

20 at 100%	2,000
20 at 80%	1,600
40	3,600

Average 90%

Loss of MPP due to labor market...... 10%

Summary of all MPP losses due to adverse conditions:

Labor market	10%
Material deliveries	2%
Working conditions	8%
Total loss	20%

The total loss is 20%, and the net MPP is 80% of normal. To bring normal labor units up to a point where they would be great enough to absorb all losses, a 1.25 multiplier would have to be used. Conduit that normally required 10 hr per 100 ft for installation would be figured at 12.50 hr per 100 ft, and so on for all other items if the loss were general.

In some cases the estimator figures his labor as normal and applies his additions to the total normal hours. Take a project having 1,000 hr estimated labor, and make separate additions for labor market and job conditions. The adjusted labor would be arrived at as follows:

	Hours
Estimated normal labor.....................	1,000
Add for labor market (80% MPP) (25%).....	250
Add for job conditions (10%)...............	100
Total estimated hours required............	1,350

A STUDY OF REPORTS

Before leaving this chapter there should be some comments on misleading reports. When all mechanics are employed or conditions in the building industry are hectic, we get reports about the mechanics "lying down on the job." One should not accept these reports too freely.

During World War II, reports were current that MPP was down 30%. The author studied some of the cases, and the 30% was accounted for about as follows:

Cause of Loss	Per Cent
Delayed and untimely delivery of materials.......	7
Lack of cooperation of other trades..............	5
General confusion on job.......................	3
Labor market.................................	15
Total.....................................	30

The regular mechanics were plugging along as usual but were hampered with obstacles beyond their control.

Another government wartime project reported productivity 50% below normal. The principal cause was poor management on the part of government representatives. The contract was cost-plus, and the contractor was constantly pressed to put more men on the job. As a result, the project was overmanned.

On this same job, much excess material was delivered before the actual installation work was started. Much of it was handled and protected throughout the life of the contract and then loaded and shipped away. Other difficulties, similar to those described for former contracts, prevailed. A careful study of conditions would more than likely have revealed that not more than 20% of the low productivity was a fault of the labor market.

CHAPTER 27

Advertising

The author is not an advertising man and does not intend to suggest fantastic ideas which are supposed to triple contractors' sales overnight. This chapter is for the purpose of reviewing the experiences of some successful contractors and answering questions often asked.

Not all contractors agree on the results obtained by various forms of advertising. Naturally, we may expect that some readers have had experiences at variance with those to be related here.

The following are forms of advertising selected for study:

1. Direct contact
2. Trucks (advertisements on owner's trucks)
3. Newspapers
4. Brochures
5. Employing experts
6. Miscellaneous

There are forms of entertainment that are often charged to advertising which are sales aids but, strictly speaking, are not advertising. Some of these will be noted later.

DIRECT CONTACT

Contacting potential buyers directly has been considered the best form of advertising since the first use of electricity. It has been over

70 years since the first electric bell men (they were not called electricians then) started writing their names, business, and address on slips of paper and handing it to persons that answered their knock at the door. This method of advertising has changed little.

Today, the electrical contractor sends a card or letter to the prospective buyer or makes a personal call. The latter is the better. With a personal call, the contractor makes sure that his message reaches the right person and he also learns whether or not it is advisable to spend any more time on the lead. The call may offer an opportunity to do some groundwork. And certainly the buyer will remember the call much longer than he would a letter.

Personal calls for actual selling have been discussed in previous chapters.

TRUCKS

Opinions vary regarding the value of the contractor's name on a truck as a source of advertising. However, we do know that the number of trucks owned by electrical contractors would be less if some of them were not convinced that there is an advertising value.

Some contractors have enough trucks to take care of several crews. Wherever the men are working, there is a truck out in front to advertise the fact.

Some of us are of the opinion that numerous trucks on the street bearing the names of electrical contractors help combat the tendency of owners to buy materials directly from supply houses and have their own men install it. This opinion was sprouted in the early 1920s.

In the early 1900s, electrical contracting was evidenced by horse-drawn trucks bearing contractors' names. With the introduction of motor trucks, contractors turned to public hauling and their names disappeared from the streets. The display of supply trucks with brightly painted names was much in evidence.

At the same time, supply-house salesmen made a drive to induce owners to buy material directly from them and have their (the owner's) men install it or give contractors installation-only orders. It is believed that the display of supply trucks and the disappearance of contractors' names from the streets did much to aid this drive to get business away from electrical contractors.

One thing we are sure of: A name on a truck is not an expensive form of advertising.

NEWSPAPERS

When we think of advertising, newspapers come to our mind. Many wonder why electrical contractors do not use them. In large cities where contractors would like to have their names introduced to more people, the rates are high and the results to be hoped for are small. In a city like Chicago, one may expect to pay $1,000 or more for a quarter-page advertisement. Sunday rates run approximately 50% higher.

In the smaller communities, rates prove more inviting but the need of getting the name before the local people is not so great.

Advertising an electrical contracting service is not like advertising a soap or food product that practically every reader is in the market for. Besides, most large buyers and many small ones have agents to act for them and are not interested in such newspaper advertisements. A large majority of the contractor's work is obtained from architects, engineers, or general contractors. All these people are contacted directly.

The lack of newspaper advertising by electrical contractors indicates that they do not consider it a good investment. However, one cannot close the door on any form of publicity. Each contractor must study his own position and community.

Some readers may ask why contractors do not advertise in papers as a group. Such a move would result principally in promoting greater use of electricity and the wiring to accommodate it. This would be more or less a duplicate of what the utilities are doing in most communities.

BROCHURES

Periodically, some contractor comes out with a brochure. Most of the brochures are elaborate, with pictures of completed projects, lists of important customers, pictures of buildings, pictures of office force, and possibly pictures of the mechanical crews. Often pictures of the original location and the present company-owned buildings are placed together to show the growth.

One company included a picture of one of its original construction crews. This picture certainly provided an interesting and attractive feature. The mechanics, with their "handle-bar" mustaches and stiff derby hats, looked very formal. Along with this picture of the original crew was a picture of all the contractor's existing crews together. The contrast no doubt made an impression on buyers.

A Sales Aid

Brochures are for old established contractors with much completed work to advertise. In reality, they are used more as sales aids than a means of advertising. The pictures, list of completed projects, personnel, and history of the organization are all designed to promote buyers' confidence.

Objections

For the ordinary contractor with a limited amount of business, the following are the objections to the brochure:

1. It takes a great deal of some valuable employee's time for the selection of materials and the direction of the preparation.
2. The number that can be used effectively is limited.
3. It is soon obsolete.

EMPLOYING EXPERTS

Evidently the big advertising firms do not consider the electrical contracting industry too good a field for their business because they do not give it much attention.

A national association once retained a reputable advertising firm to make recommendations and suggest methods to be used by its members. The recommendations were made before a meeting of leading members of the association.

This meeting was held in the big ballroom of a leading hotel. The walls were lined with great pictures like a giant diorama encircling the entire room. These pictures, together with suitable captions, were to be reproduced in brochures which were to be distributed by the contractors to potential buyers.

The project was a very costly one. Temporarily, there was much enthusiasm, but the meeting and its resultant brochures were soon history. The association wanted to find out what experts could do for its members, and the only thing to do was to hire one and pay the price.

MISCELLANEOUS

The individual contractors have a variety of ideas for advertising. Some have letterheads with special captions. Others have letterheads with lists of completed projects or pictures of buildings they have wired.

At times, contractors have open house. Such affairs must be man-

FIG. 27-1. A page from a 32-page brochure by Newbery Electric Corp., Los Angeles, Calif. Other pages provided additional information of interest to prospective buyers.

aged carefully. It is better to have nothing than to have customers feel that they have wasted an evening. An excess of liquor can quickly give a party a "black eye."

The author recently attended an affair given by a contractor engaged in a wide variety of electrical-construction work. The evening opened with a buffet supper. At each place was a favor consisting of a pen-and-pencil set. After the supper, there was a conducted tour of the plant, during which the use of special construction equipment was demonstrated. A feature of the party was that each of the contractor's representatives was given a list of guests to look after. It was a wholesome affair and no doubt paid dividends.

There are numerous forms of advertising on a small scale such as special Christmas cards, Christmas favors, and useful gadgets with the firm's name on them. Considering the cost, the results are worthwhile.

We can agree that a contractor's best advertisements are his completed installations, but he must do something to procure the installations. Completed projects help hold existing customers, but one must advertise to get new accounts. For this purpose an expense allowance must be made.

EXPENSE ALLOWANCE FOR ADVERTISING

Electrical Estimating lists 0.25% of business volume (cost) as a reasonable allowance for advertising. This is generally ample if confined to advertising as outlined in the foregoing paragraphs. The 0.25% represents the average expense. The amount may vary greatly from year to year.

Many contractors report expenses greatly in excess of the recommended amount. We learned in Chap. 10 that such reports must be studied carefully.

Some contractors include country club dues and expenses, entertaining, convention trips, and many other expenses under the heading of advertising. In the first place, any such items chargeable to business are promotional expenses separate from those we have covered. In the second place, much of such expense should be charged to the owner's personal account.

In discussing problems of electrical contracting, we always come back to the same statement: "Each business has its individual requirements and management must act accordingly." Before one attempts to emulate others, he must be sure that he understands what they are doing and that he is operating under similar conditions.

CHAPTER 28

Continuing in Business

In previous chapters we have been interested in building up and conducting a business that could endure. Before closing, it is well that certain cautions be viewed.

After a business is started and running on a sound basis, the most serious hazards are eliminated. However, there is never a time when a man in business can relax his hold and operate without being on his guard. In electrical contracting, the two principal hazards seem to be impatience and depressions. Among the less frequent causes of trouble can be listed such items as careless spending, ostentation, overconfidence, apathy toward employees, and sickness.

IMPATIENCE

Regardless of how carefully a contractor plans his work, there are likely to be times when there is a temporary slump in contracts. His regular customers have no work, and he has not been successful in competitive bidding. His work is running low, and he has good mechanics that he does not want to let go.

A contractor in such a position has a real problem, and we cannot sit at a desk without having all the details and prescribe a solution. If the slump lasts too long, it is hard for many contractors to be patient and keep from getting panicky. There are many decisions that must be carefully weighed.

The contractor must decide how much it is worth to keep his organization intact. He must also decide whether it is better to take out-of-the-pocket money to keep men on the payroll or to take contracts below cost. It often takes much study for a contractor to decide on the best program to follow.

In general, it is found better to take several small jobs at a low price than it is to take one large contract below cost. While the small contracts are being completed, the contractor has a chance to study what lies ahead and can plan his program accordingly. However, the tendency is to want to take one large contract. The gambling spirit makes men want to play for the bigger stakes. There is always the hope that on the larger jobs there will be additional orders and, on the job, savings which will help pull the job out of the hole. Besides, men get impatient and do not want to creep along with small jobs; they want to get one large contract and have the worry of holding mechanics out of the way.

Again, contractors get impatient and panicky when bidding. After losing several jobs, they start cutting prices until they get a "loser." They have let the other fellow set the price. It cannot be stated too often that one of the surest ways for a contractor to get headed for bankruptcy is to let his competitors do his estimating.

A contractor gets a large project at a very low price. The labor is figured low because his best men are available. After construction is under way, his regular customers have work and want the mechanics that have always served them. The contractor has no choice. He must take the much needed mechanics off the large project and hire whomever he can get for the work. So the bad job is still worse off.

We have witnessed no end of cases where contractors became impatient and took large contracts at a loss. Most of them survived the shock, but the experience put an awful kink in their bankroll and left a blot on their credit. It takes patience and good management to weather a slump. Using out-of-the-pocket money to keep business intact is painful, but it is not so serious as heavy losses on large contracts.

Another serious result of impatience comes from contractors trying to grow big too quickly. They take large contracts at low prices and get so loaded up with work that their organization is spread out too thinly to function well. The results range from minor losses to total bankruptcy.

DEPRESSIONS

Business goes in cycles. There are boom times and slumps. For a very long period cycles were of approximately 10 years' duration with

7 good and 3 bad years. In the past 25 years the pattern has varied. Regardless of the cycles, the electrical contractor can expect, sooner or later, to experience a serious slump in business that lasts for a long period. The smart operators try to prepare to withstand the strain that such times put on them.

Three important steps a contractor can take to be prepared for a depression are:

1. Build up a reserve
2. Avoid long-term high-cost leases
3. Avoid overexpansion

As we study them it will become obvious that these three items are more or less tied together.

Building up a Reserve

Electrical Estimating (McGraw-Hill Book Company, Inc.) gives 0.75% as the minimum to be added to the base cost of contracts for building up a necessary reserve. This amount was recommended for large volumes. The small operator requires from 2 to 5%.

The small operator is inclined to think that a prolonged slump in business will be less serious for him than it will for big business. Such is not the case. The big company can greatly reduce its personnel and still have a skeleton organization, whereas the little man already has a very limited number of employees. Take one phase of the business—bookkeeping. A large company may have three bookkeepers. If contracts drop to 25% of normal, it may be able to get along with one and his salary could be reduced. The owner of a one-man business has one bookkeeper. If he wants to keep his organization intact, he must keep the bookkeeper on and the best that can be done is to reduce the rate of pay. There are many other ways in which the big organization can, in proportion, reduce costs more easily than a one-man business.

Required Reserve $18,000

A one-man organization would require a reserve of approximately $18,000 to keep his organization substantially intact during three bad years. We shall establish the figures first and then study possible deductions.

In Chap. 3, we studied the costs of operating a one-man business. We shall list those same figures as "normal costs" and with them show possible deductions that might be made during dull times or depressions.

OPERATING COSTS PER MONTH
For a One-man Business

Item of Expense	Normal Cost	% Reduced	Reduced Cost Amount
Proprietor's salary..................	$ 700	50	$350.00
Bookkeeper and office man...........	350	30	245.00
Rents (heat included)..............	150	20 *	120.00
Light...........................	10	25	7.50
Telephone.......................	15	20	12.00
Depreciation on equipment.........	15	50	7.50
Interest on investment.............	15	20	12.00
Automobile and travel expense......	85	50	42.50
Insurances and miscellaneous.......	15	20	12.00
Totals......................	$1,355		$808.50

Estimated reduced cost per month $808.50.

* The contractor may have a long-term lease and cannot effect any saving on rent.

The estimated reduced cost of operation is $808 per month. There is one more item of expense to be considered: the cost of carrying key men when there is no work. As we shall see later this expense may be avoided, but for the time being we shall use $100 per month. At $3 per hour, $100 represents approximately one man-day per week for a month.

Adding the $100 to the $808, we have a total of $908 as the estimated reduced operating cost per month. The amount per year is $10,896.

To pursue our problem, let us assume that in a 10-year cycle the contractor has 7 good and 3 bad years. In Chap. 4 we learned that the business we are studying would require $100,000 in contracts to supply the minimum operating costs. This $100,000 was sell price. The base cost (purchase price of material and payroll) was $75,000.

If the contracts received in the bad years amounted to 25% of those for the good years, the base cost per year would be 25% of $75,000, or $18,750.

With an annual volume (base cost) of $18,750 and an over-all markup of 25%, the contractor would have $4,687 (25% of $18,750) income to pay operating costs. Our figures showed that he would be spending $10,896. The difference is $6,209, which represents the amount he would have to take out of his reserve. During the 3 bad years the total drain on the reserve bank account would be $18,627.

The amount included in the $18,627 for holding key men was $100 per month for 3 years, making a total of $3,600. Suppose conditions were such that key men would be glad to stay on and take what work they could get. The $3,600 could be deducted, and the revised estimated cost for the 3 years would be $15,027.

THREE PER CENT FOR RESERVE

The $15,027 that was established as the reserve required to keep the business intact during the 3 bad years represents almost 3% of the base cost of all the contracts for the 7 good years. A review of the figures is as follows:

For the one-man business that we are studying, we learned (see Chap. 4) that the annual volume of business required was $100,000. This was sell price. The base cost was $75,000.

During the 7 good years the total volume (base cost) would amount to 7 times $75,000, or $525,000. The $15,027 estimated reserve is approximately 2.9% of this amount. For our work, we shall use 3%. It must be remembered that nothing was included for holding key men.

What do all the foregoing figures mean? They mean that during the 7 good years, the entire volume of business (base cost) must carry a 3% burden to build up a reserve for the 3 bad years. If conditions were favorable, the amount might be less; if unfavorable, it could be much more. Regardless of the exact figures, we cannot ignore the fact that during good years, business must carry a burden in the anticipation of critical times.

PRIDE IN THE BUSINESS

Contractors put their heart and soul, as well as their money, into the building up of a business. With some it is an avocation as well as a vocation. None want to see the results of their efforts stop bearing fruit.

Young men want their business to carry on because it is their livelihood, and old men want to see their work carried on by others because they are proud of it.

Electrical contractors complain at times, the same as all others, but in the final analysis, they enjoy their work, are proud of it, and have no desire to see it terminated.

Definitions

Adjustment Factor—Overhead. Multiplier used in estimating overhead to compensate for excess expense (overhead) during off years.

Base Cost. Combined cost of material (purchase price) and labor (job payroll).

Classification of Operating Costs. Operating costs are classified under two headings, namely, direct job costs and overhead.

Complete Installation Contract. A type of contract in which the contractor supplies the complete installation including all materials, labor, and installation services.

Complete Job Cost. Base cost plus all operating costs, i.e., prime cost plus overhead.

Common Markup. A markup that is applied to both material and labor.

Consumed or Expendable Tools. Tools such as drills, files, hacksaw blades, taps, etc., which are completely consumed on the individual job.

Cost-plus Contract. A type of contract in which material and labor are billed at base cost plus a markup for installation services (operating costs) and a service return.

Cycle Man-power Demands. The condition under which the crew is built up to the maximum and then dropped off on the completion of the job.

Depreciated Tools. Tools such as stocks, wrenches, pipe benders, and trucks which are normally depreciated only on a single job.

Direct Job Costs. Costs of operating, such as estimating, engineering supervision, tools, etc., which can be identified with a particular job.

Extended Duration. Time beyond the anticipated or contract duration.

Fixed-fee Contract. A type of cost-plus contract in which the markup is a fixed amount (lump sum).

Fluctuating Man-power Demands. The number of men required to satisfy the demands of the job fluctuations.

General Contractor. A contractor who undertakes, directly or through others, to hire and superintend all trades.

Incidental Labor. The labor consumed by preparation, study time, supervision, etc., which is not applied directly to the installation.

Installation-only Contract. A contract to supply labor, management, and installation services for the installation of materials furnished by others.

Job Costs. Direct job costs and overhead.

Job Duration. The time interval between the starting and the nominal completion of a project.

Job Factor. A multiplier used to adjust labor costs to suit the particular job being figured.

Labor and Material Contract. A type of contract in which labor, material, and job expenses are billed in detail. This is commonly referred to as a "time-material" contract.

Man-power Demands (MPD). The man-power requirements of a project, such as the average and maximum number of men and the uniformity of the demands.

Man-power Efficiency. The ability and willingness of mechanics to produce. It is the ratio of work done to average normal accomplishments.

Material Service. The service required for the proper supply and protection of materials, such as purchasing, coordinating deliveries, checking, and protecting.

Material Services—Material by Others. Services rendered in connection with materials furnished by others.

Nonproductive Labor. A common term used when referring to incidental labor.

Obsolete. Of no more economical use because it is superseded by something better.

Operating Costs. All costs, over and above base costs, required to deliver a complete installation.

Optimum Duration. The duration which provides for the most economical installation.

Overhead. Items of expense, such as rent, bookkeeping, light, telephone, and administrative, which cannot be identified with or assigned to any particular project.

Prime Contractor. One who contracts directly with the owner.

Prime Cost (Base Job Cost). Combined cost of material, labor, and direct job costs.

Procurement Failures. Failure on the part of procurement agents to supply materials in the proper quantities at the right time and in the right place.

Productive Labor. A term used to designate the labor consumed in the actual installation of the work.

Productivity—Man Power. The ratio of the actual man-hours consumed on a job to standard man-hours for similar projects.

Definitions

Profit. A permanent gain which can be taken out of the business.

Prolonged Duration. Duration out of proportion to normal.

Rated Life of Tools. The period of time that a tool may remain available for economical use.

Reserve for Contingencies and Guarantee. The amount included in the estimated cost to cover unforeseen expenses and to make good the guarantee.

Separate Markup. A markup applied to one division only (either material or labor) of the base costs.

Service Return. A markup on the complete job cost for services rendered. It is anticipated profit.

Useful Life of Tools. The actual time that a tool may be available for active and economical use.

Abbreviations

CECA	Chicago Electrical Contractors' Assn.
CEEA	Chicago Electrical Estimators' Assn.
CIC	Complete installation contract
CMU	Common markup
COH	Common overhead
CPC	Cost-plus contract
DJC	Direct job cost
DJE	Direct job expense
DL	Direct labor
FFC	Fixed-fee contract
FMPD	Fluctuating man-power demand
GC	General contractor
IO	Installation only
IOC	Installation-only contract
IL	Incidental labor
L&M	Labor and material
LMR	Labor-material ratio
LMU	Labor markup
LO	Labor only
LOH	Labor overhead
MPD	Man-power demand
MOH	Material overhead
MPE	Man-power efficiency
MPP	Man-power productivity
MU	Markup
NECA	National Electrical Contractors' Assn.
NPL	Nonproductive labor
OH	Overhead
PC	Prime contractor
PL	Productive labor
UP	Unit price

Index

Abbreviations, 279
Adjustment factors, labor units, 260
   unbalanced ratios (M/L), 98
Advertising, 263–268
  expense allowance, 268
  forms of, 263
    brochures, 265
    direct contact, 263
    employing experts, 266
    miscellaneous, 266
    newspapers, 265
    trucks, 264
Alteration work, 61–63
Analysis, of incidental labor, 177
  of labor costs, 153
  of management, 209
  of operating costs, 81
  of overhead, 89
Apprentice training, formal, 193–200
    cost of, 193–199
    division of cost, 194
    excess carrying cost, 198
    practice materials, 198
    scope of training, 194
    total cost, 198
    value of services, 196
  informal, 42
Apprentices, billing for time, 112
  expense of training, 193
  productive efficiency, 196
  wages, 200
Architects and engineers, 69, 237
  educational work with, 242
  maintaining good will, 244
  protecting, 243
  protecting the owner, 243
  recommending changes in plans, 242
  specified markups for bidding, 240

Architects and engineers, unit price bidding, 238
  varied markups for bidding, 240

Base costs, 82
  divisions of, 82
Bid spread, 221–228
  causes of, 221
    condition of business, 221
    contractors' resourcefulness, 228
    faulty markups, 226
    rushed estimates, 228
  studying low bids, 228
Bidding, 119
Billing and collecting, comprehensive, 229
  for contracts, 141
    cost-plus, large, 123–125, 150
      small, 141–144
    fixed-fee, 150–151
    fixed-price, large, 144–150
      small, 141
  labor, 186, 190–191
  markups, specified, 240
    varied, 240–242
  regular buyers, 230–234
  safe billing, 229–236
  time and material orders, 144
  unit prices, 238–240
Business, continuing in, 269–273
  cost of starting, 9–21, 27
    office equipment and supplies, 11
    stock, 15
    storeroom, 12
    tools, 12
    total, 21
    trucks, 12
    (*See also* Operating costs)

Business, failures, 4
  volume required, 29

Closed shop, 42
Collecting, 141–152
  billing for (see Billing)
Common multiplier, 239–240
Complete job costs, 82
Completed-job studies, 171–173
Concealed hazards, 62
Continuing in business, 269–273
  business hazards, depressions, 270
  impatience, 269
  pride in business, 273
  reserve required, 271–273
Contracting, electrical, competition, 5
  confidence required, 7
  cost of (see Cost; Costs)
  failures, 4
  functions of, 49–55
  as good business, 6
  hazards, 3
  volume required, 29
Contracts, obtaining, 33–39
  selecting, 61–69
Cost, of starting business, 9–21, 27
  of waiting for contracts, 23–27
Cost-plus contracts, bidding and collecting for, 123–125, 141–144, 150
  definition, 279
  installation-only projects, 111–114
Cost studies, labor, 153–175
  overhead, 89–101
Costs, advertising, 268
  base costs, 82
    labor, 82
    material, 82
  direct job, 82, 89–101
  incidental, 177–192
  operating (see Operating costs)
  overhead, 29, 82, 89–101
Curves, 209–214
  labor, 210, 211, 213
  man-power demands, 209
  management analysis, 209
  as medium of expression, 209
  overhead, 91–93

Dangerous practices, 120
Definitions, 275–277
Desirable contracts, 64
Direct job costs, 29, 82
  estimating of, 89–101
  and incidental labor, 180–184
  installation-only contracts, 109
  listing, 186–187
  percentage breakdown, 185
Drawings, installation, 52–59
  job, 73–75

Efficiency, man-power (see Productivity)
Engineering, 51–52
Estimating, 49–51
Estimator, 247–251
  contentment of, 251
  freedom of activity for, 248
  overtime and overtime pay for, 249
  responsibility of, 248
Excess labor, billing for, 113
  contracts, 108
  installation-only contracts, 108
Expense (see Cost; Costs)

Failures, business, 4
Fixed-fee contract, billing for, 114, 150–151
  definition of, 275
  for installation-only contract, 114–117
  management of, 114
Fixed-price contract, 122–125, 141
  billing and collecting for, 141, 144–150
Functions of electrical contracting, 49–55
  directing work, 71, 73
  engineering, 51, 73
  estimating, 49
  figuring operating costs, 81
  selecting better work, 61
  selecting mechanics, 71
  selling, 119–132

General contractor, 69

Hazards, of alteration work, 61
  of business, 3
  concealed, 62
  general, 65
  of general contractor, 69
  in installation-only contracts, 104–106
  of occupied spaces, 62
  procurement failures, 77
  progress of other trades, 62

Incidental costs, 177–192
  detailed, 189–192
  not overhead, 180
Incidental labor, 177, 180–182
  check list for, 180–181
  and direct job cost, 180–184
  included in standard labor unit, 178
Individual unit method, 238–239
Industrial projects, 63
  advantages of, 63
  lead to future work, 64
  mechanics required, 63
  quick turnover of, 63
Inferior mechanics, cost of, 44

Index 283

Installation-only contracts, 103–117
  cost-plus form, 111–114
  definition of, 103
  direct job expense of, 109
  excess labor required, 108
    chart for, 105
  fixed-fee contract, 114–117
  hazards of, 104–106
  requirements of, 103
  selling of, 110
Insurance subject to low markup, 185

Job costs, 82
  complete, 82
  direct, 82
  effect of labor management on, 78–79
  overhead, 82
Job-division studies, 159–167
Job studies, 90

Labor, billing for, 186
  burden, 36
  cost studies, 153–175
    adjustment of labor units, 167
    benefits from, 153, 174
    breakdown studies, 155–156
    complete projects, 171–173
    contracts completed below estimated cost, 173–174
    incidental labor, 180
    job divisions, 159–167
    program of procedure, 154
  directing men, 73
  inferior mechanics, 44
  installation-only contracts, 108, 113
  and job costs, 78–79
  man-power problems, 41–47
  market, 253–258
  miscellaneous, 178, 180
  nonproductive, 177–178
  open shop mechanics, 42
  and operating costs, 36–38
  productivity of, 253
    crew, 258
    individual, 254–258
  selecting mechanics, 71–73
  trim labor, 215–220
  union mechanics, 43
  working conditions, 62, 259–260
Labor-only, term erroneously used, 103
Labor units, adjusting, 159, 167–171, 260
  standard, 178–179
Lien, waiver of, 151–152
Limited partnership, 138
Loss of customers, 235

Man-power, problems of, 41–47
  inferior mechanics, 44

Man-power, problems of, labor market, 253
  open shop mechanics, 42
  seniority board, 42, 43
  productivity, 253–261
Management, charts of, 78–79
  curves show effect of, 210
  effect on trim labor, 216
Markups, 31
  faulty, 226–227
  specified, 240
  varied, 240–242
Material, check list for, 20
  delivery of, 77
Material-labor ratio, 67, 98, 103
Material service, 99
  installation-only contracts, 109
  estimating, 108, 109, 114
  selling, 109, 110
Mechanics (*see* Labor)

Nonproductive labor, a misnomer, 177
  nil on well-managed jobs, 178

Office, equipment required, 11
  location, 19
  rents, 19
One-man business, cost of starting, 9
  (*See also* Operating costs)
  division of proprietor's time, 10
  expense of office, 11
  miscellaneous expenses, 12, 22
  volume of business required, 23
Open shop mechanics, 42
Operating costs, 23–27, 29–32, 81–88
  definition of terms, 82
  direct job costs in, 82
  for one-man business, established, 30–32
    first 6 months, 25
    first year, 26
    per month, 24
  and overhead, 82
  studies of, 84–88
Orders, verbal, 131
Organization, business, 133–140
  corporation, 138–140
    advantages of, 139
    obtaining charter, 138–139
  limited partnership, 138
  partnership, 135–138
    advantages of, 135–136
    disadvantages of, 136–138
    phases to be studied, 135
  sole owner, 133–135
    advantages and disadvantages of, 134–135
Overhead, 82
  adjustment for unbalanced M/L ratios, 98
  analysis of, 89
  complete studies of, 95

Overhead, curves, $100 to $9,000, 91
  $10,000 to $100,000, 92
  $100,000 to $1,000,000, 93
 errors in estimating, 89
 studies of, 89–101
  complete, 95
  material and labor (table), 86
  preliminary, 90
  various types of work (table), 88
 surveys of, 84
 value of limited studies, 100
Overmanning work, 172

Pipe bender, hydraulic, 16
Plans and working drawings, 52–59, 73–75
Platform hoist, 17
Productivity, man-power, 253
  adjusting labor units for, 260
  for a crew, 258
  factors affecting, 253
   labor market, 253
   management, 259
   working conditions, 259
  for individual, 254–258
Profit, 83
 use of *return* instead, 110
Proposals for electrical work, 119–132
 completion dates supplied with, 122
 dangerous practices in submitting, 120
 samples of, 120, 121, 124
 special case, 125–128
 types of, 120–122
  cost-plus, 123–125
  fixed-price, according to plans, 120–121
   no plans, 122–123
  verbal orders, 131

Quick turnover in industrial work, 63

Ratios, material/labor, 67
 study of, 107
 unbalanced, 98
Reasons for going into business, 2
Rents, office, 19
Reserve required for business, 271–273
Return, 110

Saw, band, 16
Selecting, contracts, 61
 mechanics, 71

Selling, 110
Seniority board, 43
Stock requirements, new business, 15–19
Storeroom, 12
Strikes as incentive to starting business, 4
Studies, apprentice training, 193–199
 labor costs, 153–175
 operating costs, 81, 84–94
 overhead, 89–101
Substantiating charges, 186, 189
Surveys, business, 84

Tools, billing for, 207
 cost of, 5-man crew, 13, 204, 206
  50-man crew, 204–205
 explaining charges for, 201
 factors affecting costs, 202
 individual job expense, 203
 inventory rating of, 4
 life of, 203–204
 required, for 5-man crew, 13
  for 50-man crew, 205
 time in use vs. rated life, 202
Training of apprentices (*see* Apprentice training)
Trim labor, 215–220
 faulty analysis of, 216–219
 management of, 216
 mechanics, "letdown," 219
 pickup work, 217
 psychological influence, 220
 reducing crews for, 216
Trucks, advertising value, 264
 expense of, 12

Union mechanics, 42
 hiring, 43
  seniority board, 43

Verbal orders, 131
Volume of business required, 29

Wage rates, 23–24
Waiver of lien, 151–152
Working conditions, 259–260
 alteration work, 62
 occupied spaces, 62